Practical ECG
FOR
Exercise Science
AND
Sports Medicine

Greg Whyte, PhD, FACSM

Liverpool John Moores University

Sanjay Sharma, BSc, MD, FRCP, FESC

St George's University of London

Human Kinetics

Library of Congress Cataloging-in-Publication Data

Whyte, Gregory P.
  Practical ECG for exercise science and sports medicine / Greg Whyte, Sanjay Sharma.
    p. cm.
  Includes bibliographical references and index.
  ISBN-13: 978-0-7360-8194-8 (soft cover)
  ISBN-10: 0-7360-8194-1 (soft cover)
  1. Electrocardiography. 2. Exercise tests. 3. Heart--Diseases--Diagnosis. 4. Exercise--Physiological
aspects. I. Sharma, Sanjay. II. Title.
  RC683.5.E5W49 2010
  616.1'207547--dc22

                    2010011261

ISBN: 978-0-7360-8194-8 (print)

The Web addresses cited in this text were current as of March 12, 2010 unless otherwise noted.

Acquisitions Editors: John Dickinson, Loarn D. Robertson, PhD, and Karalynn Thomson; **Developmental Editor:** Kevin Matz; **Assistant Editors:** Casey A. Gentis and Steven Calderwood; **Copyeditor:** Patsy Fortney; **Indexer:** Bobbi Swanson; **Permission Manager:** Dalene Reeder; **Graphic Designer:** Fred Starbird; **Graphic Artist:** Yvonne Griffith; **Cover Designer:** Bob Reuther; **Photographer (interior):** Photos courtesy of Greg Whyte, unless otherwise noted; **Photo Production Manager:** Jason Allen; **Art Manager:** Kelly Hendren **Associate Art Manager:** Alan L. Wilborn; **Printer:** Total Printing Systems

Printed in the United States of America   10 9 8 7 6 5 4 3 2

The paper in this book is certified under a sustainable forestry program.

Human Kinetics
Web site: www.HumanKinetics.com

United States: Human Kinetics
P.O. Box 5076
Champaign, IL 61825-5076
800-747-4457
e-mail: humank@hkusa.com

Canada: Human Kinetics
475 Devonshire Road Unit 100
Windsor, ON N8Y 2L5
800-465-7301 (in Canada only)
e-mail: info@hkcanada.com

Europe: Human Kinetics
107 Bradford Road
Stanningley
Leeds LS28 6AT, United Kingdom
+44 (0) 113 255 5665
e-mail: hk@hkeurope.com

Australia: Human Kinetics
57A Price Avenue
Lower Mitcham, South Australia 5062
08 8372 0999
e-mail: info@hkaustralia.com

New Zealand: Human Kinetics
P.O. Box 80
Torrens Park, South Australia 5062
0800 222 062
e-mail: info@hknewzealand.com

For Seema, Anushka, and Shivani

Sanjay Sharma

To my wife, Penny, and my precious girls,
Maya and Elise

Greg Whyte

Contents

Part III

The Athlete's Heart

Preface

The electrocardiogram (ECG) was first described by Einthoven (a Dutch physician) in the late 19th century. Following the invention of the bipolar limb leads by Einthoven, Dr. Frank N. Wilson and colleagues established the six chest leads in the 1930s. That was followed by Dr. Goldberger's invention of the three augmented limb leads. In the subsequent years, ECG has become the most widely used noninvasive tool in the assessment of cardiac electrical function. The ECG offers the opportunity to examine the heart to discover information about the electrical conduction system and its morphology, function, and circulation. *Practical ECG for Exercise Science and Sports Medicine* is based on our experience evidenced by over 100 peer-reviewed papers, multiple books and book chapters, and over a decade of interpreting ECGs from clinical patients to elite athletes both at rest and during exercise.

Practical ECG for Exercise Science and Sports Medicine offers an overview for those unfamiliar with the ECG and a resource for those working in the field of exercise and sports medicine and science. Featuring more than 100 ECGs and illustrations, the seven chapters in this text progress from theory to practice. Chapter 1 presents an overview of the morphology of both the heart and the electrical conduction system. An examination of cardiac function at rest and during exercise offers the underlying theory relevant to the interpretation of the ECG. Chapter 2 examines methods of monitoring heart rate and rhythm, facilitating an understanding of the breadth of measurement modalities available. Chapters 3, 4, and 5 examine the ECG in detail at rest and during exercise. Chapter 3 focuses on obtaining a 12-lead ECG, including

troubleshooting common problems. Chapter 4 offers an interrogation of normal and abnormal ECG findings with a detailed examination of cardiac abnormalities. Each abnormality is accompanied by a sample ECG trace. Chapter 5 focuses on the ECG during exercise. Exercise creates a unique physiologic situation that results in a provocative challenge to the cardiovascular system. As a result, the ECG during exercise has become a tool in the identification of a variety of pathologies, including coronary artery disease, exercise-induced arrhythmias, and assistance in the differentiation of pathologic and physiologic enlargement of the left ventricle. While the role of the exercise ECG for the identification of myocardial ischemia has been in use for decades, a burgeoning understanding of the importance of the exercise ECG in identifying and differentiating various diseases has resulted in an increased use of exercise as a diagnostic tool in cardiology. Chapter 6 focuses on athletes' hearts. The physiologic adaptations associated with chronic physical training manifest as cardiac enlargement and increased thickness of the walls, additional heart sounds on auscultation, sinus bradycardia, and ECG anomalies. These physiologic changes may mimic those observed in pathologic processes, so accurate diagnosis is required for differentiating a physiologic from a pathologic substrate. This is of particular importance for those pathologies that cause sudden cardiac death in athletes. An understanding of the anomalies often present in athletic individuals at rest and during exercise is important for those working in sporting environments. Chapter 7 presents case studies to highlight the problems encountered in dealing with athletes' hearts.

Understanding a normal ECG at rest and during exercise and being able to interpret findings are becoming increasingly important for noncardiologists, including health professionals, sports medicine specialists, physiotherapists, clinical exercise physiologists, and sport and exercise scientists. Accordingly, *Practical*

ECG for Exercise Science and Sports Medicine contains timely information for undergraduate and postgraduate students from a variety of disciplines. In addition, this book is a valuable resource for health professionals working with athletic and sedentary individuals at rest and in exercise stress testing.

STRUCTURE AND FUNCTION OF THE HEART

Part I provides an introduction to the heart and a foundation for understanding the ECG. Chapter 1 presents an overview of the anatomy of the heart and morphology of both the heart and the electrical conduction system. Also, an examination of cardiac function at rest and during exercise offers the underlying theory relevant to the interpretation of the ECG. Chapter 2 then examines methods of monitoring heart rate and rhythm, facilitating an understanding of the breadth of measurement modalities available. Part II will then begin to cover the ECG in detail at rest and during exercise. Part III focuses specifically on the athlete's heart.

1

The Heart

The heart is a four-chamber muscular pump located in the mediastinum within the thoracic cavity. The left side of the heart (left atrium and left ventricle) is responsible for pumping blood around the systemic circulation, and the right side of the heart (right atrium and right ventricle) is responsible for pumping blood around the pulmonary circulation. The heart is held within a thin fibrous sac called the pericardium and is composed of three layers: epicardium, myocardium, and endocardium. The myocardium is composed of a network of cardiac muscle cells that are branched and abut end-to-end allowing the transmission of an action potential and resultant contraction arising at a single point to spread across the entire myocardium (functional syncytium).

During the cardiac cycle electrical, mechanical, and valvular events are coordinated to produce two distinct phases: systole (contraction) and diastole (relaxation). The action potential that signals the commencement of cardiac contraction is initiated in the pacemaker cells of the sinoatrial node (SA node) and spreads across the atria in a wavelike action leading to atrial depolarization (systole) indicated by a P wave on a surface electrocardiogram (ECG). Reaching the AV junction, the action potential is conducted along the bundle of His and through the right and left bundle branches and Purkinje's system resulting in depolarization of the ventricular myocardium (ventricular systole) represented by the QRS wave on the ECG. Ventricular repolarization is represented by the T wave and completes the characteristic P-QRS-T nomenclature of the surface ECG. The translation of electrical activity to myocardial contraction is characterized by length–tension, force–velocity, and pressure–volume relationships.

In response to exercise, changes in the neural control and the hormonal milieu result in alterations in the chronotropic (heart rate) and inotropic (contractile force) state of the heart. The impact of these changes results in an increased cardiac output as a result of an increased heart rate and stroke volume associated with an increased myocardial oxygen demand in the normal heart.

CARDIAC ANATOMY

The heart is a muscular pump approximately the size of a closed fist situated in the mediastinum within the thoracic cavity and orientated to the left of the sternum. The heart is shaped like a blunt cone. The superior flat end of the cone is termed the base, and the inferior rounded point of the cone is termed the apex. The base is directed posteriorly toward the right shoulder, and the apex is directed anteriorly toward the left hip and can be palpated in the fifth intercostal space. The right ventricle faces the anterior chest wall.

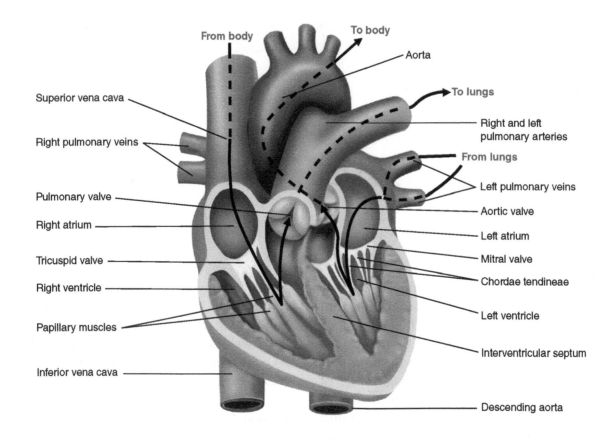

Figure 1.1 Chambers of the heart.

Reprinted, by permission, from J. Wilmore, D. Costill and W.L. Kenney, 2008, *Physiology of sport and exercise*, 4th ed. (Champaign, IL: Human Kinetics), 125.

The heart consists of four chambers divided into two pumps (see figure 1.1). The right atrium receives relatively deoxygenated blood from the systemic circulation via the superior and inferior vena cava flowing through the tricuspid valve into the right ventricle, which pumps blood to the lungs through the pulmonary valve and into the pulmonary artery (pulmonary circulation). The left atrium receives blood from the pulmonary circulation via the four pulmonary veins flowing through the mitral valve into the left ventricle, which pumps blood to the systemic circulation through the aortic valve and into the aorta.

A nonpermeable wall that consists of the interatrial septum and interventricular septum separates the left and right sides of the heart. Rings of connective tissue, the atrioventricular

(AV) rings, separate the atria from the ventricles and act as a fibrous skeleton for the origin and insertion of atrial and ventricular muscle and for the attachment of the bases of the mitral, tricuspid, aortic, and pulmonary valves.

The heart is held within a thin fibrous sac called the pericardium and is composed of three layers: the epicardium, myocardium, and endocardium (see figure 1.2). A space between the pericardium and the epicardium contains a small volume of pericardial fluid that acts as a lubricant. The epicardium, lined by a parietal layer of mesothelium, covers the myocardium (cardiac muscle) and is relatively stiff preventing excessive acute enlargement of the heart. The endocardium lines the inner surface of the myocardium and is composed of endothelial cells and is a continuation of the endothelium of blood vessels.

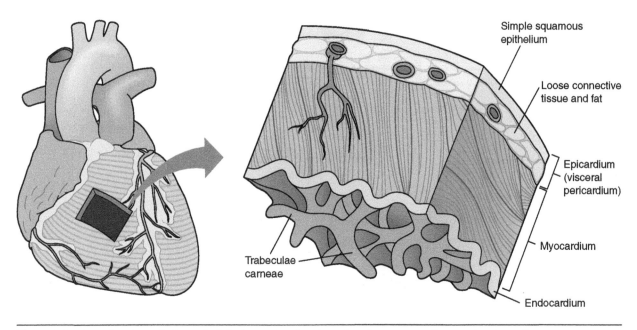

Figure 1.2 A section through the wall of the heart showing the epicardium, myocardium, and endocardium.

A large coronary sulcus (the groove along which the coronary arteries run) runs around the heart at the level of the atrioventricular ring with a further two sulci extending inferiorly from the coronary sulcus. The anterior interventricular sulcus and the posterior interventricular sulcus indicate the division between the right and left ventricles.

The right and left coronary arteries are the major arteries supplying blood to the tissues of the heart originating at the left and right sinus of Valsalva found in the ascending aorta just above the point where the aorta leaves the heart. The right coronary artery is smaller than the left and supplies blood to a smaller portion of the total cardiac tissue. The path of the right coronary artery follows the coronary sulcus from the aorta to the posterior part of the heart. The branches of the right coronary artery supply blood to the lateral wall of the right ventricle, and a branch called the posterior interventricular artery follows the posterior interventricular sulcus and supplies blood to the posterior and inferior part of the heart. The left coronary artery descends along the anterior interventricular sulcus

and bifurcates into the anterior descending (anterior interventricular) artery and the circumflex artery. The anterior descending artery supplies blood to most of the anterior part of the heart. The circumflex artery follows the coronary sulcus to the posterior side of the heart where it supplies blood to the posterior wall. A marginal branch of the circumflex artery supplies blood to the lateral wall of the left ventricle (see figure 1.3).

The great cardiac vein drains the left side of the heart, and the small cardiac vein drains the right side of the heart. Converging in the posterior coronary sulcus, the two cardiac veins empty into the coronary sinus, which in turn empties into the right atrium. A number of smaller veins empty directly into the coronary sinus or the right atrium.

The myocardium is composed of cardiac muscle cells. Similarities exist between cardiac and skeletal muscle cells including the arrangement of actin and myosin filaments into sarcomeres with A, I, and H bands and M and Z lines. In contrast, however, cardiac muscle cells are significantly shorter and narrower than those of skeletal muscle.

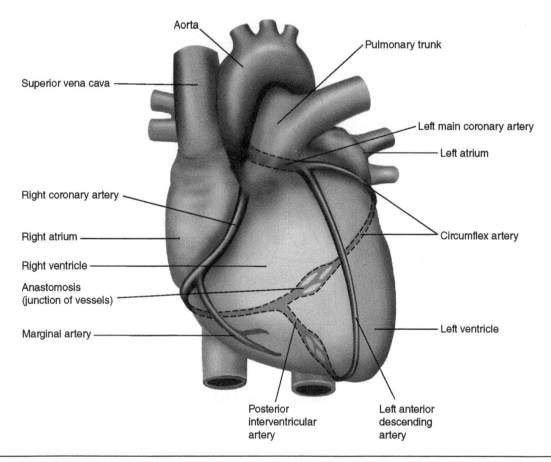

Figure 1.3   Circulation of blood to the heart.

Reprinted, by permission, from J. Wilmore, D. Costill and W.L. Kenney, 2008, *Physiology of sport and exercise,* 4th ed. (Champaign, IL: Human Kinetics), 127.

Furthermore, cardiac muscle cells are branched, abut end-to-end to form a network (syncytium), and have a higher mitochondrial density. In addition, each cardiac muscle cell has a single nucleus. The sarcoplasmic reticulum of cardiac muscle is less extensive than that of skeletal muscle, with wider T tubules located at the Z line. The interface between two cardiac muscle cells occurs at one of the Z lines in the form of a specialized dense intercalated disc. A low-resistance electrical pathway is created within these intercalated discs forming a gap that allows an electrical current to pass between adjacent cells. Within the discs, desmosomes provide sites of adhesion between cells allowing the transfer of developed tension from one cell to the next. These structures form a functional syncytium characteristic of cardiac muscle allowing the transmission of an action potential and resultant contraction arising at a single point to spread across the entire myocardium (functional syncytium).

CARDIAC FUNCTION

The cardiac cycle refers to the relationships among electrical, mechanical, and valvular events during one complete heart beat (see figure 1.4). At a resting heart rate of 70 beats·min[-1], the cardiac cycle lasts 0.85 s and is divided into two distinct phases: systole (contraction) and diastole (relaxation). At rest, ventricular systole is about 0.3 s, and diastole is 0.55 s. At a heart rate of ~200 beats·min[-1], however, systole is 0.15 s and diastole is 0.15 s for a total cardiac cycle length of 0.3 s.

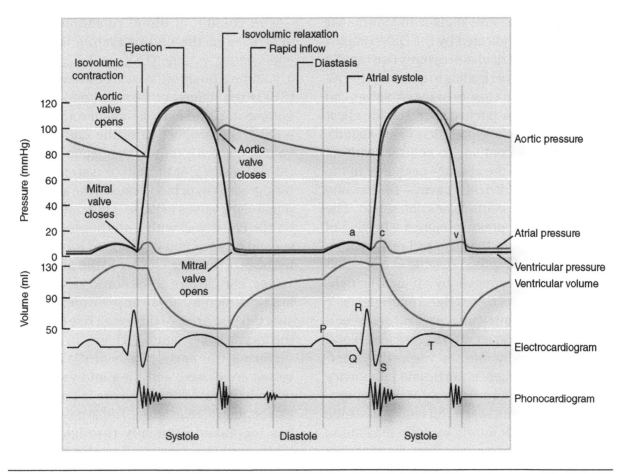

Figure 1.4   Electrical, mechanical, and valvular events within the left ventricle during the cardiac cycle.

Figure 14.27, p. 433 from HUMAN PHYSIOLOGY, 2ND ED. BY Dee Unglaub Silverthorn. Copyright © 2001 Prentice-Hall, Inc. Reprinted by permission of Pearson Education, Inc.

The sequence of events is similar but slightly asynchronous on the right and left sides of the heart. The electrical conduction across the atria results in the right atrium contracting before the left atrium. In the ventricles, the left ventricle contracts first, although blood is ejected from the right ventricle prior to the left as a result of the significantly lower pulmonary artery pressure. The duration of the right ventricular ejection is longer than that of the left because of the lower pulmonary pressures, resulting in a delayed closure of the pulmonary valve.

In middiastole, the point just prior to atrial systole, the atrial and ventricular pressures are low with atrial pressures slightly higher than ventricular pressures leading to the AV valves being open. As the AV valves open, blood flows into the ventricles in the absence of atrial systole (this is termed the early phase of filling). The ventricles contain ~80% of their final volume prior to the depolarization of the atria indicated by the P wave on the electrocardiogram (ECG; see later for a more detailed description). Toward the end of the P wave, the atria contract increasing atrial pressure (a wave; see figure 1.4) and propelling blood into the ventricles resulting in the final 20% of ventricular filling and a small rise in ventricular pressure. The resultant ventricular volume is termed end-diastolic volume (EDV) and is about 130 ml in a standing position rising to 160 ml in a supine position at rest.

Ventricular diastole ends with ventricular depolarization indicated by the QRS complex on the ECG. Ventricular pressures rise rapidly during early systole leading to closure of the AV valves. The rapid closure of the AV valves can be heard as a low-pitched sound on the chest wall using a stethoscope (first heart sound). The ventricles continue to contract with all valves closed in the absence of change in the volume of the ventricular cavity—the isovolumetric phase of ventricular contraction. The high ventricular pressure causes a bulging of the AV valves into the atria and results in an increase in atrial pressure—the C wave (left atrial pressure increases by ~10 mmHg; right atrial pressure increases by ~5 mmHg).

When the ventricular pressures exceed aortic and pulmonary artery pressures, blood is ejected rapidly into the respective arteries. As a consequence, aortic and pulmonary artery pressures rise closely following ventricular pressures (aorta: 80 mmHg diastolic minimum to 120 mmHg systolic maximum; pulmonary artery: 8 mmHg diastolic minimum to 25 mmHg systolic maximum). The shortening of the ventricles toward the AV ring causes a lengthening of the atria increasing atrial volume and reducing atrial pressure (the X descent).

The appearance of the T wave on the ECG during midsystole indicates repolarization of the ventricles, and toward the end of the T wave the ventricular myocardium begins to relax leading to a fall in ventricular pressures. The momentum of the blood within the left ventricle imparted during the rapid onset of contraction in early systole leads to the slow ejection of blood even when left ventricular pressure falls below aortic pressure. As a result of the resistance to flow offered by the peripheral arterial system, only around half of the blood ejected from the left ventricle is propelled through the aorta; the remaining half is accommodated by the elastic distension of the aorta until the elastic recoil of the aorta propels it into the systemic circulation.

The volume of blood remaining in the ventricles at the end of systole is termed end-systolic volume (ESV) and is ~60 ml in a standing position at rest. Stroke volume (SV) is the difference between end-diastolic volume and end-systolic volume (SV = EDV − ESV); the proportion of the end-diastolic volume ejected is termed the ejection fraction (EF = SV / EDV). At the end of systole, a small retrograde flow of blood causes the aortic and pulmonary valves to close. This can be heard on the chest wall with a stethoscope as the second heart sound (this sound is often split because of the later closure of the pulmonary valve). The closure of the aortic valve results in a small increase in aortic pressure; the pressure dip just prior to this small rise is termed the incisura (see figure 1.4).

During the early phase of diastole, all valves are closed, and the rapid ventricular relaxation and falling pressure is termed the isovolumetric relaxation. The atrial pressure has increased gradually throughout systole reaching a peak of ~5 mmHg in the left atrium and ~2 mmHg in the right atrium—the V wave, until ventricular pressure falls below atrial pressure when the AV valves open and blood flows into the ventricles. The flow of blood from atria to ventricles results in a fall in atrial pressure—the Y descent. During this early diastolic phase, the ventricles fill rapidly, leading to vibrations that may be detectable as the third heart sound. Atrial and ventricular pressures then gradually increase into middiastole, during which time the ventricles continue to fill slowly, a period termed diastasis.

ELECTRICAL PROPERTIES OF THE HEART

All myocardial cells demonstrate myogenic automaticity, the ability to contract rhythmically without nervous input. This crucial characteristic of cardiac muscle cells allows the denervated heart to continue beating in an orderly fashion. The intrinsic atrial and ventricular rates are slow (60 beats·min⁻¹ and 40 beats·min⁻¹, respectively. Ventricular rate could fall as low as 20 beats·min⁻¹ when the electrical impulse originates in the most distal Purkinje cells).

The action potential that signals the commencement of cardiac contraction is initiated in the pacemaker cells of the sinoatrial node (SA node), a specialist group of cells in the upper right atrium near the entrance of the superior vena cava. These specialist cells have a resting membrane potential that slowly and spontaneously depolarizes to a threshold at which an action potential is initiated. The pacemaker firing rate, and therefore heart rate, is increased by sympathetic nerve activity and decreased through parasympathetic activity. The autonomic nervous system influences conduction velocity and the duration of the cardiac potential. Cardiac action potentials have a long duration (200-400 ms) involving the movement of sodium (Na^+), potassium (K^+), and calcium (Ca^{2+}). In contrast to skeletal muscle contraction, cardiac muscle contraction cannot summate because of the overlap in time between the cardiac action potential and the initiated contraction.

Conduction System Anatomy

Following the depolarization of the SA node, the action potential is conducted along the plasma membrane from one atrial cell to another through the low-electrical-resistance intercalated discs. The conduction velocity across the atria is enhanced through anterior, middle, and posterior internodal bands leading to an almost simultaneous contraction of both atria. The action potential reaches the atrioventricular node (AV node) located in the fibrous AV ring on the right of the atrial septum (the AV ring is nonconducting, thus establishing the AV node as the single point of action potential conduction from atria to

ventricles). The conduction velocity of the AV node is very slow (0.05 m·s⁻¹) compared with the myocardium (0.5 m·s⁻¹) delaying the transmission of the action potential by around 0.1 s and ensuring the end of atrial contraction prior to the commencement of ventricular contraction.

From the AV node the action potential continues its journey down the ventricular septum through the bundle of His (conduction velocity of 1 m·s⁻¹), along the right and left bundle branches and into the terminal Purkinje network (conduction velocity of 5 m·s⁻¹). This network radiates through the ventricular myocardium reaching all areas of the myocardium and resulting in a near simultaneous contraction of both ventricles (see figure 1.5 on p. 10).

Basics of Cardiac Action Potentials

Cardiac potentials have three distinct patterns found in (1) the atrial myocardium, (2) the SA and AV nodes, and (3) Purkinje fibers and the ventricular myocardium. The duration of action potentials also differs: the shortest duration is in the SA and AV nodes and atrial myocardium (200-250 ms), followed by the bundle of His and ventricular myocardium (250-300 ms) and the Purkinje fibers (300-400 ms) (see figure 1.6 on p. 10).

The resting membrane potential of ventricular muscle is steady at about –80 mV. The action potential has an initial rapid depolarization (phase 0) as a result of a sudden influx of Na^+ as fast Na^+ channels open together with the efflux of K^+. The action potential reaches a peak of about +40 mV followed by a rapid but short decline associated with the inactivation of Na^+ channels (phase 1). The delayed opening of the Ca^{2+} channels together with the slow closure of some Na^+ channels leads to a prolonged plateau of the action potential (phase 2). The rapid influx of K^+ and the return of Na^+ and Ca^{2+} to normal leads to the repolarization

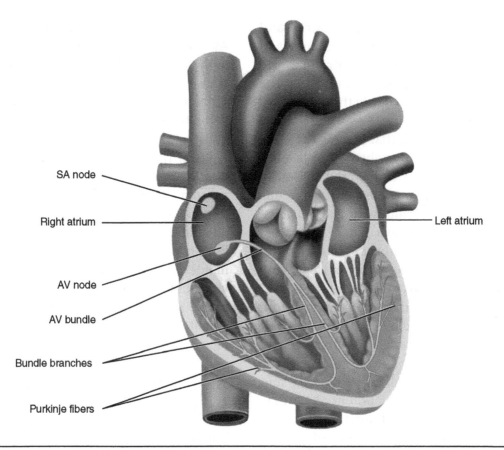

Figure 1.5   The cardiac electrical conduction pathway.

Reprinted, by permission, from J. Wilmore, D. Costill and W.L. Kenney, 2008, *Physiology of sport and exercise,* 4th ed. (Champaign, IL: Human Kinetics), 128.

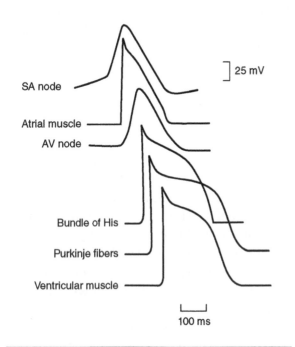

Figure 1.6   The shape, duration, and sequence of cardiac action potentials.

Reprinted from B. Hoffman and P. Cranfield, 1960, *Electrophysiology of the heart* (New York: McGraw-Hill), 261.

of the membrane in phase 3. In atrial muscle, the action potential follows the same ionic mechanisms as the ventricles with minor differences in the time course of events. A less distinct plateau (phase 2) results in a longer repolarization phase (phase 3).

The cells of the SA and AV nodes have a less negative resting potential as a result of a high membrane Na+. Between action potentials their resting potential (phase 4) is less stable because of a gradual fall in K+ leading to a slow, spontaneous depolarization from –65 mV to around –50 mV. This characteristic is called the prepotential, or pacemaker potential. The action potential is initiated when the threshold potential of –50 mV is reached. The lack of fast Na+ channels in nodal cells leads to a slow depolarization (phase 0) reaching a peak of about +20 mV. The nodal cells repolarize at a rate similar to that of the atrial myocardium. Because of the steeper slope of the pacemaker potential in the SA node, it is the SA node that triggers the action potential that spreads through the atria, causing the atrial action potential, which arrives at the AV node before the pacemaker in the AV node has time to reach its threshold. In this way the SA node initiates the action potential and dictates the rate of action potentials that are conducted through the AV node, bundle of His, and ventricular myocardium (i.e., heart rate; see figure 1.6).

The cardiac cell is not excitable during the absolute refractory period of the action potential (~200 ms). Excitability gradually recovers during the subsequent relative refractory period (~50 ms), and a second action potential can be elicited with a progressively decreasing strength of action potential throughout the relative refractory period. An action potential generated during and immediately after the relative refractory period has a slower rate of depolarization with a characteristically low amplitude and short duration. There is a latency of ~10 ms from the start of the action potential before the myocardium begins to contract; a peak in developed tension occurs just prior to the end of the absolute refractory period.

By the end of the relative refractory period, the cardiac muscle cell is halfway through its relaxation phase. It is therefore impossible for cardiac muscle to produce summation and tetanus because of the overlap of electrical and mechanical events. The prolonged cardiac refractory period is of critical physiological importance: it protects the ventricles from too rapid a reexcitation, thus allowing adequate filling time.

Autonomic Control of Cardiac Electrical Activity

The control of action potential frequency (heart rate) and duration is termed chronotropy and is normally controlled by the cardiovascular centers of the brain via the autonomic nervous system and the hormone adrenaline. The sympathetic nerves increase heart rate, and the parasympathetic nerves decrease heart rate. The normal activity of the parasympathetic nerves is often termed vagal tone because the fibers project to the heart along the vagus nerve. At rest, both sympathetic and parasympathetic nerves are active; however, the parasympathetic activity is dominant resulting in a resting heart rate of ~70 beats·min^{-1}. A heart rate below 60 beats·min^{-1}, termed bradycardia, is normally associated with an increase in vagal tone and a decrease in (and possible absence of) sympathetic drive. In contrast, heart rates above 100 beats·min^{-1}, termed tachycardia, occur as a result of an increased sympathetic tone and vagal withdrawal.

Noradrenaline (norepinephrine) and adrenaline (epinephrine) bind to plasma membrane beta-adrenoceptors in the SA node increasing the rate of pacemaker potential depolarization. The earlier action potential triggering allows a greater number per unit time. In contrast, acetylcholine released from parasympathetic fibers acts on the muscarinic cholinoceptors in the SA node to slow the rate of pacemaker potential depolarization. Sympathetic and parasympathetic fibers also innervate the atria and AV node, and the sympathetic fibers innervate the Purkinje fibers and ventricular myocardium altering the conduction velocity in the expected manner. These changes allow normal cardiac function at high heart rates by shortening all phases of the cardiac cycle.

The upper limit to heart rate is dictated by the minimum action potential duration of the atria and the conduction rate of the AV node. Although the maximum rate of atrial contraction is ~400 beats·min^{-1} (minimum action potential ~120 ms), the AV node is unable to conduct more than 220 action potentials per minute. Therefore, a coordinated heart is unable to beat faster than 220 beats·min^{-1}.

ELECTROCARDIOGRAPHY

Electrodes placed on the skin can detect the depolarization and repolarization of the myocardium, which can then be displayed on an electrocardiograph and printed in the form of an electrocardiogram (ECG). In two dimensions the amplitude (vertical displacement) of the ECG signal represents the magnitude of the electrical signal propagated by the myocardium (i.e., the summation of electrical signals from all myocardial cells at any time point). The amplitude is, therefore, a representation of the myocardial mass; the larger the muscle mass, the larger the amplitude (vertical displacement) of the ECG signal measured in millivolts (mV). The duration (horizontal distance) of the ECG signal represents the duration of the electrical event and is measured in seconds (for ease of interpretation, duration is often reported in millimeters, mm).

Commencing the cardiac cycle midway through diastole, the cardiac myocardium is electrically stable and represented on the ECG as a flat line termed the isoelectric line. In the normal heart the depolarization process is initiated at the SA node leading to the depolarization of the entire atrial myocardium (described earlier). This process is represented by a small vertical displacement of the ECG from the isoelectric line represented by the P wave on the ECG. The P wave has a small vertical displacement as a result of the low myocardial mass of the atria. The initial phase of the P wave reflects right atrial depolarization, whereas the later phase reflects left atrial depolarization.

The action potential then reaches the non-conducting AV ring, where the signal is terminated in all cells except the specialist cells of the AV node, which conduct it across the AV junction along the bundle of His, left and right bundle branches, and Purkinje network prior to the initiation of ventricular depolarization. The conduction velocity through the AV junction (AV node and bundle of His) is slow leading to a period of electrical inactivity and a return of the ECG to the isoelectric line. The period of time between the initiation of atrial depolarization and the start of ventricular depolarization is represented by the PR interval. Ventricular depolarization causes a vertical displacement of the ECG that is larger than that observed for atrial depolarization because of the greater muscle mass of the ventricles (particularly that of the left ventricle).

The rapid conduction of the action potential results in a rapid depolarization of the entire ventricle myocardium represented by the QRS complex on the ECG. The repolarization of the ventricles is represented by the ST segment, T wave, and U wave (a U wave is not always present). Atrial repolarization is not seen in the normal heart because of the very low-amplitude signal and the presence of the dominant QRS signal (atrial repolarization may be seen in certain pathological conditions such as pericarditis; see chapter 4). Following the repolarization of the ventricles, the ECG returns to the isoelectric line prior to the commencement of the next cardiac cycle. In summary, the electrical activity of the cardiac cycle is represented on the ECG by the P-QRS-ST-T-U sequence (see figure 1.7).

CONTRACTILE PROPERTIES OF THE HEART

The contractile process of the heart is similar to that of skeletal muscle, except that the activation of the contractile process through a rise in cytosolic free- Ca^{2+} is mediated by both intra- and extracellular Ca^{2+} in the heart. During the depolarization phase the contractile process begins with the entry of Ca^{2+} along voltage-sensitive channels in the sarcolemma and T tubules leading to an increase in Ca^{2+} concentration from 10^{-7} molL^{-1} to 10^{-6} molL^{-1} during systole. Ca^{2+} entering via the sarcolemma partially activates the contractile process and also acts as a trigger for the release of Ca^{2+} from the sarcoplasmic reticulum (Ca^{2+}-induced Ca^{2+} release). Additional Ca^{2+} enters

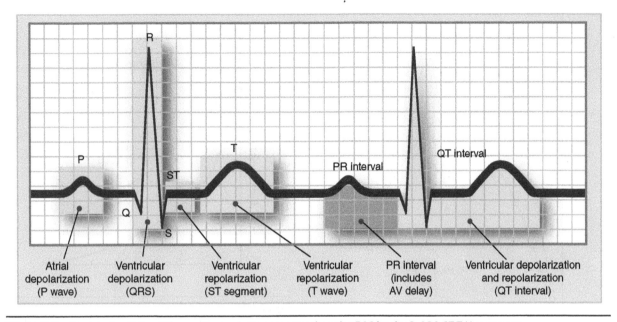

Figure 1.7 Electrical activity of the cardiac cycle is represented on the ECG by the P-QRS-ST-T-U sequence.

Reprinted, by permission, from J. Wilmore, D. Costill and W.L. Kenney, 2008, *Physiology of sport and exercise*, 4th ed. (Champaign, IL: Human Kinetics), 161.

along the Na–Ca^{2+} exchanger. Repolarization results in the reduction of intracellular Ca^{2+} concentration via the active accumulation of Ca^{2+} by the sarcoplasmic reticulum (Ca^{2+}–ATPase pump) and via the plasma membrane Ca^{2+}–ATPase pump and Ca^{2+}/Na$^+$ exchange. The availability of Ca^{2+} for the subsequent contraction is limited by the slow movement of Ca^{2+} from the uptake to the release sites of the sarcoplasmic reticulum.

Length–Tension Curve

Unlike that of skeletal muscle, the resting length of cardiac muscle cells is shorter than the optimal length (see figure 1.8). Cardiac muscle cells normally operate on the ascending part of the length–tension curve, and as a consequence, any small change in initial length will result in large increases in active and total tension. This rapid increase in tension is the result of more effective cross-bridge formation and a stretch-related increase in intracellular Ca^{2+} concentration and sensitivity of troponin to Ca^{2+}. This intrinsic characteristic of cardiac muscle cells underpins the intrinsic control of stroke volume.

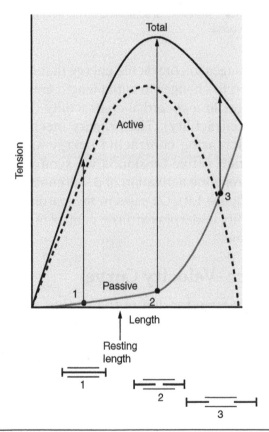

Figure 1.8 Total and active tension developed during an isometric contraction. Points 1, 2, and 3 demonstrate developed tension at three different starting cardiac muscle lengths. The resting length of cardiac muscle cells is shorter than the optimal length.

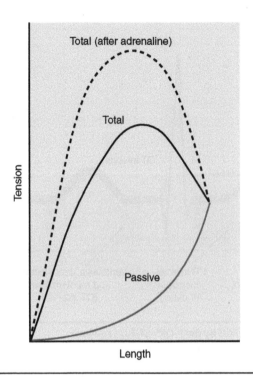

Figure 1.9 The effect of adrenaline on the length–tension curve.

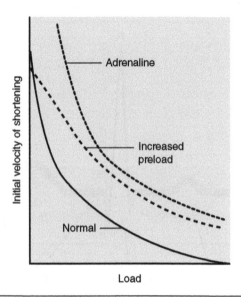

Figure 1.10 A normal force–velocity curve demonstrating a reduction in the initial velocity with an increased afterload and the effect of an increased preload or increased adrenaline on the load-velocity curve.

Changes in contractile energy that are not related to changes in initial length are called changes in myocardial contractility (inotropic contractility). The primary mechanism for increasing contractile energy and the resultant active tension is via sympathetic or adrenaline activation of β_1-adrenoceptors (see figure 1.9). Changes in myocardial contractility underpin extrinsic control of stroke volume.

Force–Velocity Curve

The passive tension of the cardiac muscle prior to contraction is termed the preload and is related to the end-diastolic volume (EDV). The afterload is associated with the load against which the heart must contract to propel blood into the systemic or pulmonary circulation. To that end, afterload is related to the pulmonary and aortic pressure. A rise in afterload results in a reduced velocity of contraction and degree of shortening. The effect of afterload on velocity is represented in a force–velocity curve (see figure 1.10). Changes in the velocity of stroke volume ejection are associated with alterations in the velocity of shortening. Velocity of shortening is increased with an increased preload, particularly at higher afterloads. Although adrenaline increases the velocity of shortening for any given afterload, the greatest effect is observed with smaller afterloads.

Pressure–Volume Curve

Whereas length–tension curves pertain to strips of cardiac muscle, the pressure–volume curve represents the heart. Length, passive tension, and total tension are now represented by ventricular volume, diastolic ventricular pressure, and maximum systolic ventricular pressure (see figure 1.11). Ventricular volume increases from around 60 ml to an EDV of around 130 ml during diastole (point A to B in figure 1.11). At EDV, ventricular pressure has increased to ~5 mmHg. The ventricular pressure rises from 5 mmHg to ~80 mmHg (depending on aortic pressure) during the isovolumetric phase of ventricular contrac-

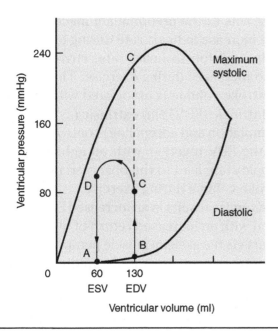

Figure 1.11 Pressure–volume curve of the left ventricle. ABCDA represents a normal cardiac cycle. EDV and ESV represent end diastolic and end systolic volume.

the contribution of atrial contraction, the magnitude of atrial and ventricular pressure during diastole, the distensibility of the ventricle, and the effect of changing intrathoracic pressure.

In contrast, an increase in afterload (aortic pressure) leads to a smaller SV. Acutely, this in turn results in an increased EDV resulting in an increased SV (Frank-Starling mechanism). This process continues until a new steady-state pressure–volume relationship has been reached. This new steady state is usually associated with a shift in the pressure–volume loop to the right and the ventricle operating at higher pressures for the same SV. The adaptive capacity of the ventricle is limited, and in the presence of large or prolonged increases in afterload, a reduction in SV will occur.

Extrinsic factors depend on an increased Ca^{2+} availability resulting in enhanced myocardial contractility and ejection velocity (inotropic contractility). The key factors influencing inotropic contractility are sympathetic stimulation and adrenaline. Increasing or decreasing stimulation results in a shift of the Starling's curve (see figure 1.12). Stroke volume can be increased in the absence of an increased EDV or maintained in the presence of an increased afterload following sympathetic stimulation.

tion (point B to C in figure 1.11). As the aortic valve opens, ventricular pressure rises to ~120 mmHg and then falls to ~100 mmHg (point C to D in figure 1.11) resulting in the ejection of ~70 ml of blood into the aorta (stroke volume). Following closure of the aortic valve, isovolumetric ventricular relaxation results in a drop in ventricular pressure (point D to A in figure 1.11). The area bounded by the pressure–volume loop ABCD represents the external work done by the heart.

The volume of blood ejected by the heart, the stroke volume (SV), is controlled by factors intrinsic and extrinsic to the heart. Intrinsic factors such as increased EDV (preload) lead to increased ventricular pressure and volume at the start of isovolumetric contraction and result in a larger pressure-volume loop and a larger SV ejected at a faster velocity. The relationship between EDV and SV is termed the Frank-Starling mechanism and is represented by the Starling's curve. EDV is affected by the time interval between contractions (i.e., heart rate), the velocity of ventricular relaxation,

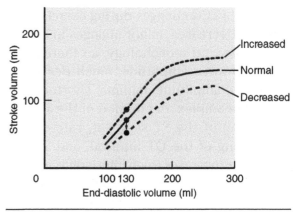

Figure 1.12 Impact of the sympathetic activity on the Starling's curve.

EXERCISE AND THE HEART

Exercise is a potent physiological stress leading to a disturbance in whole-body homeostasis. In response to exercise, changes in the neural control and the hormonal milieu result in alterations in the chronotropic and inotropic state of the heart. The impact of these changes results in increased cardiac output as a result of increased heart and stroke volume ($Q = HR \times SV$).

The increase in heart rate observed at the onset of exercise is associated with a rapid withdrawal of vagal tone. Sympathetic nerve activity begins to contribute to increases in heart rate 10 to 20 s following the onset of exercise and becomes increasingly important as heart rate increases. The heart rate has an upper limit that varies among individuals, but in general is associated with age. Maximum heart rate decreases with age at a rate of around 1 beat·year^{-1} and can be estimated using the following simple equation: HR_{max} = 220 – age. More complex calculations are available that are specific for gender and training status.

The large interindividual difference in maximum heart rate means that measurement during an exercise test to maximal volitional exhaustion remains the gold standard. The altered action potential duration, conduction velocity, and contractile velocity associated with the increase in heart rate results in a number of ECG changes during exercise: RR-interval decreases, minor changes in P-wave amplitude and morphology, an increase in septal Q-wave amplitude, small decreases in R-wave amplitude, minimal shortening of the QRS complex, depression of the J point, upsloping of the ST segment, rate-related shortening of the QT interval, and T-wave amplitude increases (high interindividual variability). There may also be a superimposition of P waves and T waves on successive beats, which indicates the shortening of diastole that acts as the predominant mechanism for the increase in heart rate during exercise.

In addition to heart rate, stroke volume also increases during exercise. The increase in stroke volume is associated with changes in intrinsic (EDV) and extrinsic (sympathetic stimulation and adrenaline) control. Changes in the EDV together with an enhanced inotropic state lead to the observed increase in stroke volume during exercise.

Exercise results in an increased EDV associated with an increased return of blood to the heart via the skeletal muscle pump, decreased venous capacitance, and rapidly changing intrathoracic pressure. The increased EDV is supported by sympathetic-induced enhancement in atrial contractility during moderate-intensity exercise. The increased EDV leads to an increased stroke volume associated with an increased length of myocardial fiber (see the preceding discussion of the Starling's curve). This intrinsic ability to increase stroke volume has an upper limit associated with the optimal myocardial length (see figure 1.13). The rapid filling of the ventricles during early diastole is important during exercise when heart rate increases and diastole is significantly reduced.

An increased sympathetic input concomitant with elevated circulating adrenaline results in an increased inotropic state (myocardial contractility). Enhanced atrial contractility increases EDV, whereas enhanced ventricular contractility decreases end-systolic volume as a result of an increased force and velocity of contraction. The result is an increased ejection fraction during exercise that results in an increased SV.

Despite a reduction in systole at higher heart rates, the rapid ejection of blood during early systole results in a minimal impact on stroke volume. However, at maximal exercise, EDV remains close to resting values despite a sympathetic-derived increase in ventricular relaxation as a result of the reduced time for

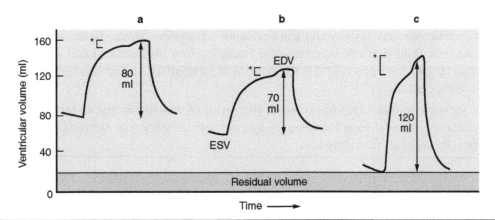

Figure 1.13 Volumes of the heart while *(a)* supine, *(b)* while standing, and *(c)* during exercise. * denotes atrial contribution to ventricular filling.

the slow phase of diastolic filling (diastasis) and the resistance of the AV valves to higher flow rates. As a result, SV rises during light and moderate exercise but reaches a plateau at intensities above 40 to 50% of maximum oxygen uptake (~120-140 beats·min⁻¹) in untrained individuals. A plateau may not be reached, and SV may continue to rise in highly trained individuals.

As myocardial oxygen demand increases (rate pressure product = systolic blood pressure × heart rate; RPP = SBP × HR), myocardial blood flow increases with increasing workload. This increase in blood flow is under local control but may also be affected by circulating adrenaline contributing to the vasodilation of coronary vessels via the activation of β_2-receptors.

KEY POINTS

▶ The heart is a four-chamber muscular pump responsible for propelling blood around the systemic (left atrium and left ventricle) and pulmonary (right atrium and right ventricle) circulation.

▶ Cardiac potentials have three distinct patterns found in (1) the atrial myocardium, (2) the SA and AV node, and (3) Purkinje fibers and the ventricular myocardium. The duration of action potential also differs; the shortest duration is in the SA and AV nodes and atrial myocardium (200-250 ms), followed by the bundle of His and ventricular myocardium (250-300 ms) and the Purkinje fibers (300-400 ms).

▶ Cardiac muscle cannot produce summation and tetanus because of the overlap of electrical and mechanical events. The prolonged cardiac refractory period is of critical physiological importance because it protects the ventricles from too rapid a reexcitation allowing adequate filling time.

▶ At a resting heart rate of 70 beats·min⁻¹, the cardiac cycle lasts 0.85 s and is divided into systole (0.3 s) and diastole (0.55 s). At a heart rate of ~200 beats·min⁻¹, the cardiac cycle is 0.3 s in length (systole is 0.15 s; diastole is 0.15 s).

▶ Electrical activity of the heart is characterized by P-QRS-T on the surface ECG, where the P wave is atrial depolarization, the QRS complex is ventricular depolarization, and the T wave is atrial repolarization.

- Chronotropy is normally controlled by the cardiovascular centers of the brain via the autonomic nervous system and the hormone adrenaline. Stroke volume is controlled by intrinsic (end-diastolic volume—the Frank-Starling mechanism) and extrinsic (inotropic) factors. Inotropy is under the control of sympathetic or adrenaline activation of β_1-adrenoceptors.

- The increase in heart rate observed at the onset of exercise is associated with a rapid withdrawal of vagal tone followed by sympathetic activity that becomes increasingly important as heart rate increases.

2

Monitoring the Electrical Activity of the Heart

Methods for monitoring the electrical activity of the heart range from the basic palpation (feeling) of the cardiac pulse wave to the highly sophisticated analysis of electrical activity using surgically implanted loop recorders. The choice of method is primarily based on the level of interrogation required to evaluate the electrical activity of the heart. Heart rate and rhythm can be determined very simply from auscultation (listening) of the heart with a stethoscope or palpation of a peripheral artery. This information is valuable in the initial assessment of a person in a clinical setting or for the rapid identification of heart rate in an exercise setting. When a more detailed interrogation of cardiac electrical activity is required, the resting 12-lead ECG is a simple, rapid, low-cost diagnostic tool. During exercise, adapted 12-lead ECG systems (stress-testing systems) allow the interrogation of cardiac electrical activity. When symptoms are intermittent, an ambulatory ECG monitor allows the evaluation of cardiac electrical activity over a prolonged period of time (up to seven days). On occasion it may be necessary to interrogate cardiac electrical activity over significantly

longer periods of time (months); in these circumstances implantable loop recorders are employed. This chapter describes the range of cardiac monitoring tools available.

PALPATION

Often, the simple palpation of a peripheral artery is sufficient to provide information about heart rate and rhythm. In response to the rapid ejection of blood from the left ventricle during systole, an increased arterial pressure results in a dilation of the aorta. Because of the peripheral arterial system's resistance to flow, only around half of the blood ejected from the left ventricle is propelled through the aorta. The remaining half is accommodated by the elastic distension of the aorta until the elastic recoil of the aorta propels it into the systemic circulation. As the blood passes through the arterial tree into the conduit arteries (i.e., carotid, subclavian, brachial, radial, and femoral arteries), a similar response to this pulsatile ejection of blood is observed with an initial dilation of the artery followed by elastic recoil to propel blood along the artery. This pressure wave, or pulse, is related to the pulse pressure (systolic blood pressure and diastolic blood pressure).

The depth of conduit arteries varies around the body. In certain places the artery is close enough to the surface to feel the pulsatile recoil of the artery to ascertain the heart rate and rhythm. Commonly used points at which these pulses can be palpated include the carotid (see figure 2.1*a*), radial (see figure 2.1*b*), tibial (see figure 2.1*c*), and occasionally femoral arteries. As blood moves into the arterioles and into the capillaries, the pulse wave is dampened and becomes almost absent. Because of this loss of the pulse in the capillaries, the fingertips can be used to palpate the pulse (see figure 2.1, *a-c*).

Counting the number of beats over a minute allows the determination of heart rate (beats·min⁻¹). Often, the number of beats over shorter periods of time is used and multiplied by a constant to obtain beats·min⁻¹ (i.e., measure the number of beats over 15 seconds and multiply by 4). Using this method, one can identify bradycardia (<60 beats·min⁻¹), tachycardia (>100 beats·min⁻¹), or a normal heart rate (60-100 beats·min⁻¹) without the use of special equipment.

Because the pulse wave palpated at the peripheral artery is a reflection of ventricular systole, it reveals both heart rate and any variations in heart rhythm. The pulse wave palpated at the peripheral artery is a reflection of ventricular systole, and to that end, in addition to identifying heart rate, any variation in the rhythm will be directly reflected in the pulse. Normal sinus rhythm results in a rhythmically consistent pattern of equally spaced ventricular systoles (pulse waves). Accordingly, the pulse wave in the peripheral arteries should display the same rhythmical profile with a constant time period between pulses. In the presence of an abnormal rhythm (arrhythmia; see chapters 3 and 4), the pattern of pulse waves varies depending on the type of arrhythmia that can be palpated in the peripheral artery.

By the simple palpation of peripheral arteries, it is possible to ascertain heart rate and rhythm without the use of complex, expensive equipment. This method reveals a great deal of information, which has resulted in its extensive use in clinical and performance environments.

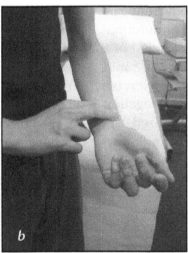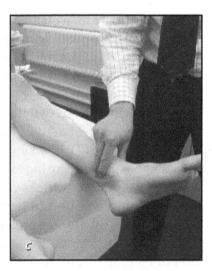

Figure 2.1 *(a)* Palpation of heart rate and rhythm at the carotid artery. *(b)* Palpation of heart rate and rhythm at the radial artery. *(c)* Palpation of heart rate and rhythm at the tibial artery.

HEART RATE MONITORS

In a sport performance environment, heart rate is commonly used to prescribe and monitor training intensities. Although palpation can be used in these environments, heart rate can only reasonably be measured during recovery periods. To overcome this limitation, companies have developed a range of heart rate monitors that can be used during exercise. The vast majority of these systems consist of a transmitter belt worn around the chest that transmits information via telemetry to a receiver, usually a watch worn on the wrist of the exerciser (see figure 2.2). Prices of heart rate monitors vary significantly depending on the functions available (e.g., stopwatch, heart rate zone alarms, calorie counters, download capabilities, and GPS). Despite the vast array of extras available, heart rate monitors offer one primary function: heart rate measurement based on an average RR interval over a set period of time (e.g., 5 s, 15 s, 30 s) using one lead (see chapter 3). Accordingly, heart rate monitors give no information on rhythm or any ECG parameter. Using the RR interval, some heart rate monitors are able to evaluate heart rate variability (HRV). Although HRV has a clinical application, this complex assessment of sympathovagal control of heart rate in a performance environment is unclear due to the small amount of research and is beyond the scope of this book.

ELECTROCARDIOGRAPHY AT REST AND DURING EXERCISE

Chapters 1 and 3 offer a detailed explanation of the electrical properties and conduction pathway of the heart and the resultant electrical trace displayed on the electrocardiogram (ECG). Here we briefly discuss the equipment and lead placement used to obtain an ECG and the preparation of the client to optimize the quality of the ECG at rest and during exercise.

The ECG at Rest

The resting ECG is the most commonly used type of ECG in the clinical setting. Therefore, it is important to have a thorough understanding of the equipment and lead placement used to obtain an ECG and the preparation of the client to optimize the quality of the ECG.

Equipment

A number of commercially available electrocardiographs (see figure 2.3 on p. 22) offer a variety of standards and unique features. The vast majority of them have a number of basic features including a high-resolution screen (a touch screen in high-end machines) displaying the ECG in real time prior to printing (saving time and paper), a high-resolution printer, automated waveform interpretation and report generation (on-screen and hard copy), ECG data storage, lead attachment alarms to warn of faulty lead connections, and

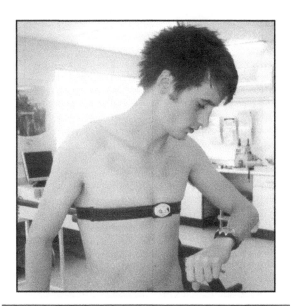

Figure 2.2   A heart rate monitor with transmitter belt and receiver watch.

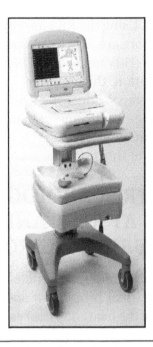

Figure 2.3 An electrocardiograph.

battery power for portability. The interpretation of waveforms provides an instant assessment of the ECG using algorithms that are specific to each manufacturer. Users should be careful about relying solely on the automated interpretations of commercial ECGs; a careful examination of the ECG should also be used. New innovations in electrocardiograph technology include wireless connection to other systems allowing the electronic transmission of ECGs.

Lead Placement

In a standard resting 12-lead ECG (figure 2.6), 10 electrodes are placed on the chest and distally on each of the limbs. Four leads are limb leads, and six leads are chest (precordial) leads. Ten electrodes generate 12 leads; the six chest electrodes generate six leads, and the four limb electrodes generate six more leads—I, II, III, aVR, aVL, aVF.

There is a universally recognized color and attachment site for each of the leads, as follows.

Limb Leads

In the resting 12-lead ECG, the limb leads are positioned on the right and left ankle and right and left wrist and are commonly assigned the following labels (note that though the photos are black and white, the color coding of the labels is universal): right leg (RL, black), left leg (LL, green), right arm (RA, red), and left arm (LA, yellow) (see figure 2.4).

Chest Leads

The chest leads are labeled V_1 to V_6 and are positioned in the following manner (see figure 2.5):

V_1—Fourth intercostal space, right sternal border (red)

V_2—Fourth intercostal space, left sternal border (yellow)

V_3—Midway between V_2 and V_4 (green)

V_4—Fifth intercostal space, midclavicular line (brown)

V_5—Midway between V_4 and V_6 (black)

V_6—Midaxillary line at the same level as V_4 (purple)

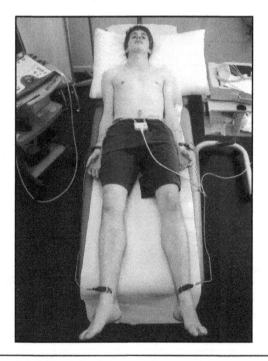

Figure 2.4 Limb lead placement.

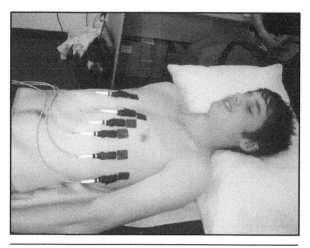

Figure 2.5   Chest lead placement.

The fourth intercostal space can be located by placing the finger at the top of the sternum and moving down the sternum to the angle of Louis, or the sternomanubrial joint (a horizontal ridge around 4 cm from the top of the sternum and adjacent to the second intercostal rib). Just below and lateral to this point is the second intercostal space. A common mistake is to use the nipple as an anatomical marker (particularly for the midclavicular line); doing so can lead to incorrect lead placement. In women and obese people the electrode may need to be placed under the breast tissue to ensure correct placement. Palpating the correct location of the fourth intercostal space on both the right and left sternal border is important because the horizontal orientation of the ribs and the level of their sternocostal attachment may vary slightly from right to left even in people with normal chest walls.

Patient Preparation: Applying Pads and Attaching Leads

This section outlines three steps to enhance the quality of the electrical signal received by the electrocardiograph, thereby enhancing the process of interpretation and diagnosis.

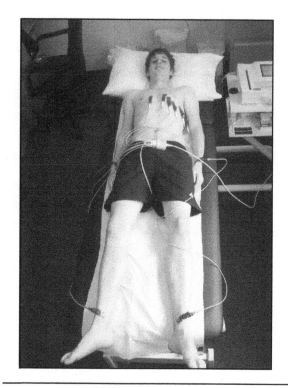

Figure 2.6   Resting 12-lead electrode placement.

- Prepare the skin using alcohol and abrasion.

- Reduce movement artefacts created by the client and the equipment.

- Reduce contact with clothing.

Step 1: Skin Preparation

Preparing the skin prior to attaching electrodes is an important step in optimizing the ECG recording. The primary purpose of skin preparation is to reduce electrical impedance. This is achieved by removing surface oils and dead skin from the electrode placement sites. First, alcohol pads should be used to remove surface oils. The skin should then be allowed to dry and then rubbed with an abrasive pad (gauze or emory paper) to remove dead skin from the surface. Prior to using alcohol, it may be necessary to shave the skin to remove excess hair (see figure 2.7 on p. 24).

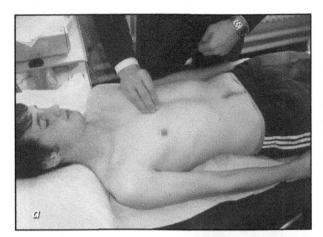
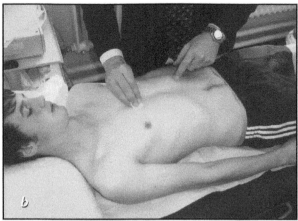
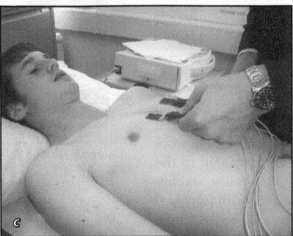

Figure 2.7 The use of *(a)* alcohol, *(b)* abrasive pads, and *(c)* lead placement to reduce electrical impedance.

Step 2: Movement Artefact

The electrical activity propagated by active skeletal muscle can cause disruption in the electrical signal received and displayed on the ECG (often termed artefact). To avoid this problem, clients should be instructed to remain motionless during the acquisition of the ECG. This may require that they place their hands under their buttocks. Providing a comfortable environment can assist in the reduction of movement artefact. The client should be placed in a supine or semirecumbant position and supported with a pillow. Heavy and deep breathing can lead to a wandering baseline (see figure 2.8) increasing the difficulty in interpreting the ECG. The client should be instructed to breathe normally and avoid deep breaths during the acquisition of the ECG.

In addition to client movement, any movement of the equipment during the acquisition of the ECG can create artefact. Careful placement of the ECG module will avoid creating tension in the leads. Leads should be untangled to avoid contact and movement between leads.

Step 3: Clothing

Contact between the electrodes and clothing can result in artefact. Clothing should not touch the electrodes during the acquisition of the ECG.

ECG During Exercise

Exercise is commonly used in the clinical setting to identify individuals with compromised myocardial blood flow or an underlying pathology predisposing to exercise and exertion

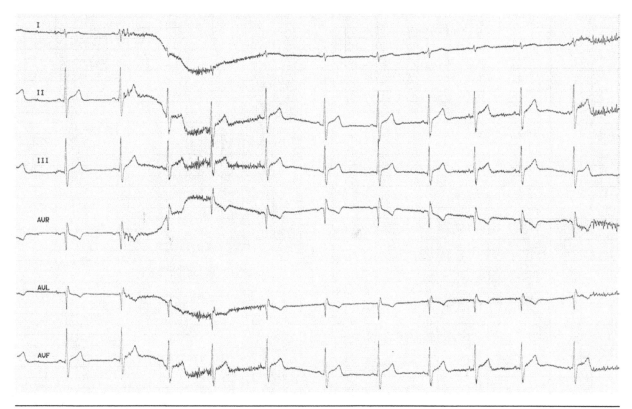

Figure 2.8   A wandering baseline associated with breathing and muscle movement. Note the artefact: the rapid, sharp, nonuniform waveform on the ECG (the wiggly line) associated with muscular contractions.

related arrhythmias. Exercise stress testing may also be used to establish the functional capacity of a patient with cardiovascular disease and evaluate the efficacy of pharmacotherapy (see chapter 5). There are a number of technical limitations in obtaining an ECG during exercise, including lead placement, movement of the patient and equipment, and the interface with ergometers (treadmill, bike, etc.) and other measurement equipment. Accordingly, specialized equipment, lead placement, and subject preparation are required to ensure optimal acquisition of ECG's.

Equipment

Resting electrocardiographs have a number of limitations in exercise environments. The two key limitations relate to the electronic filtering of the signal received by the ECG, which is used to reduce movement artefact of the equipment (leads) and the client; and the size of the monitor screen, which is often too small on resting 12-lead systems to allow optimal interrogation of the ECG waveform. Accordingly, a number of commercially available stress-testing electrocardiographs are adapted for the measurement and interpretation of the electrical activity of the heart pre-, during, and postexercise.

During exercise a significant amount of movement of both the leads and the client can cause disturbance to the electrical signal leading to artefact (noise) on the ECG and making interpretation difficult. Applying filters to the signal removes some of this artefact. Furthermore, a number of simple steps in skin preparation and lead placement can significantly reduce artefact (discussed later).

During exercise the ability to monitor the ECG waveform in real time is crucial. Accordingly, the monitor must be high resolution and large enough to allow accurate evaluation of the ECG waveform across the intensity domains (rest to maximum exercise).

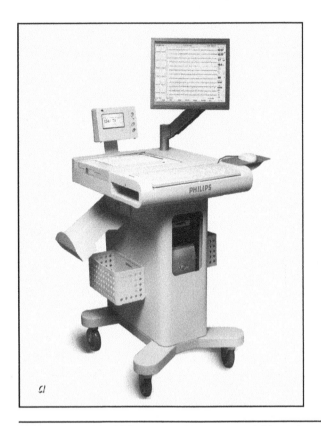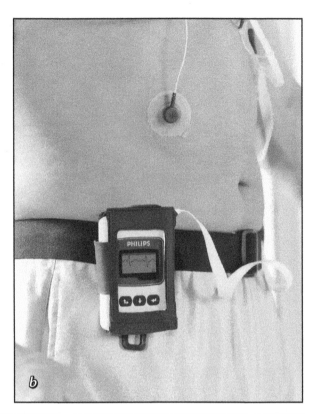

Figure 2.9 *(a)* An ECG stress-testing system and *(b)* attachment of the patient module to help reduce movement artefact.
Photo courtesy of Philips.

Standard functions offered by most ECG stress-testing systems include automated report generation and electronic storage and transmission of data and reports to improve the ease and speed of use and interpretation. New innovations include touch screen technology and wireless modules for client comfort, which helps to reduce artefact (see figure 2.9).

Lead Placement

For exercise tests the leads should be placed on clients while they are in position for the exercise (e.g., standing for treadmill testing, seated for cycle ergometer testing). The attachment sites of the chest leads during exercise are the same as during rest (detailed earlier). There are, however, distinct differences in the limb lead placement sites. Limb leads cannot be placed on the ankles and wrists during exercise because the excessive movements of these areas cause significant

artefact. Instead, the limb placement sites are moved to the torso and placed as follows (see figure 2.10):

- Right arm (RA) is placed just below the right clavicle in the midclavicular line.
- Left arm (LA) is placed just below the left clavicle in the midclavicular line.
- Right leg (RL) is placed just below the right rib cage border in the midclavicular line.
- Left leg (LL) is placed just below the left rib cage border in the midclavicular line.

The primary reason for attaching the electrodes at these sites is to reduce movement and avoid contact with clothing. In some people, particularly the obese, these sites may be highly mobile during exercise causing a significant level of artefact. To improve the quality of the ECG signal, the limb lead

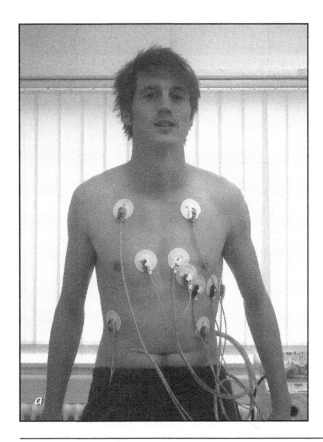
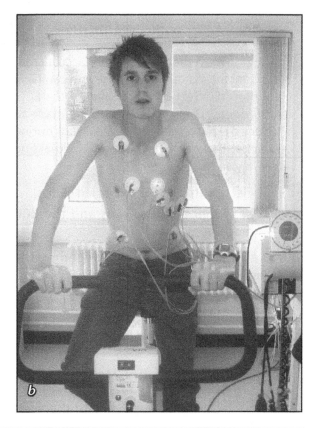

Figure 2.10 Location of limb leads during *(a)* standing and *(b)* biking exercise.

electrodes can be placed on the back of the client in the following fashion (see figure 2.11):

- Right arm (RA) is placed in the middle of the right trapezium above the scapula in the midclavicular line.
- Left arm (LA) is placed in the middle of the left trapezium above the scapula in the midclavicular line.
- Right leg (RL) is placed just below the right rib cage border in the midclavicular line.
- Left leg (LL) is placed just below the left rib cage border in the midclavicular line.

Patient Preparation: Applying Pads and Attaching Leads

As is the case with the ECG at rest, the initial preparation of the skin is imperative to optimize the quality of the ECG trace during exer-

cise. The same simple steps as those detailed for the ECG at rest should be adopted for the exercise ECG.

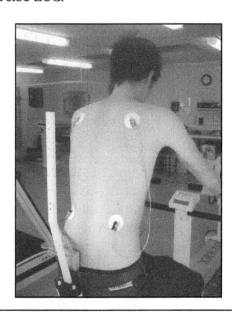

Figure 2.11 Alternative location of limb leads on the back during exercise.

Reducing Movement Artefact

Although movement artefact may occur at rest, it is guaranteed during exercise and can have a profound effect on the quality of the ECG. Accordingly, every effort should be made to reduce the movement of the ECG leads during exercise. This can be achieved by addressing two key areas:

- Electrode placement (detailed earlier)
- Movement of the leads and the ECG module

Lead movement can be reduced in the following ways:

- Use tension loops in the leads. A loop is created in the lead close to the electrode, which is then taped inferiorly to the skin (see figure 2.12).
- Secure leads together. Movement of the leads over one another should be reduced by securing (i.e. tape or clip) the leads together.
- Attach the ECG module around the client's waist using a belt. Place the module on the client's back avoiding the hips, where significant movement occurs. Carefully

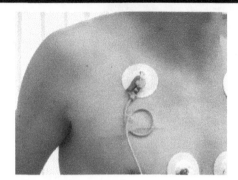

Figure 2.12   Tension loop to reduce movement artefact.

secure the lead from the ECG module to the stress-testing system to reduce movement and ensure safety (new innovations include wireless client modules that avoid this problem).

- In the presence of significant artefact from the limb leads, consider placing arm (LA and RA) electrodes at the acromioclavicular joint and placing leg electrodes (LL and RL) on the back. These alternatives cause minimal changes to the electrocardiographic picture and its interpretation.

EXERCISE MODALITIES AND PROTOCOLS

A range of exercise modalities can be used to assess the electrical activity of the heart during exercise. Standard clinical exercise tests traditionally use treadmill or cycle ergometry (see figure 2.13). Each mode of exercise has inherent positive and negative characteristics (see table 2.1 on p. 30). Accordingly, the selection of mode of exercise should account for these inherent characteristics. For example, patients with poor mobility (e.g., obese people and those with lower-limb injuries) may require upper-limb exercise modalities. In contrast, because elderly and frail people may have significant problems

with stability or with exercising to an intensity sufficient for the examination of cardiac health using a treadmill, cycle ergometry is a preferred mode of exercise in these populations. In specialist centers, other forms of ergometry may be employed including rowing and swimming (see figure 2.13). Specialized modes of exercise are valuable for those with specific symptoms; however, they require special equipment and expertise.

Before the patient starts to exercise, it is important to record a preexercise ECG in the position in which exercise is to occur (e.g., standing for treadmill exercise, sitting for cycle ergometry). Furthermore, recording a preexercise ECG during resting hyperventilation to distinguish true exertional changes

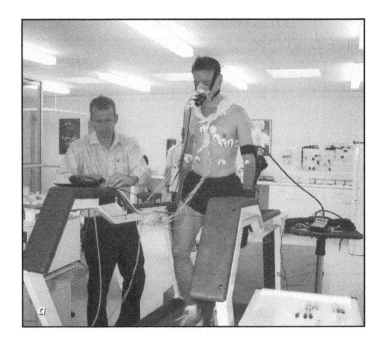

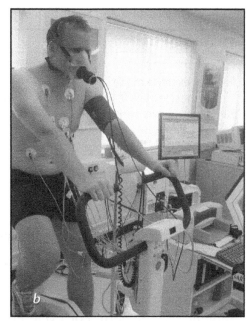

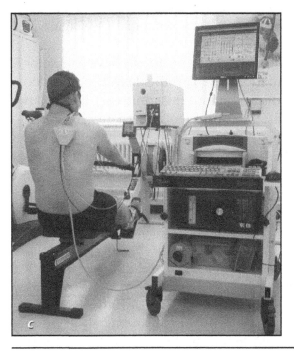

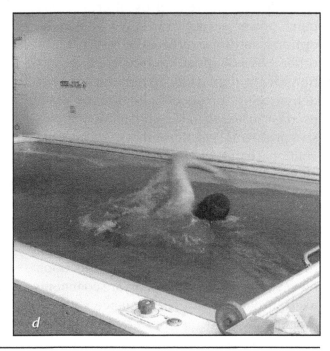

Figure 2.13 Integrated cardiopulmonary stress testing using *(a)* a treadmill, *(b)* cycle ergometry, *(c)* rowing ergometry, and *(d)* a swimming flume.

from those due simply to increased excursion of the chest wall that accompanies the increased rate and depth of ventilation may be valuable in differentiating physiological and pathological ECG changes during exercise.

The aim of an exercise stress test is to progressively increase the intensity of exercise (exercise load) to maximal volitional exhaustion (maximal exercise test) or a predetermined heart rate to bring about a (% of maximum heart rate would be the predertiminedrate in a submaximal exercise test) significant physiological stress on the cardiorespiratory system. Exercise load is increased during treadmill tests by

Table 2.1   Positive and Negative Aspects of the Use of Treadmill vs. Cycle Ergometry

Treadmill	Cycle ergometry
Positive Familiar exercise mode Larger muscle mass leads to higher $\dot{V}O_2$max High control over work rate Easy to calibrate	*Positive* Accurate quantification of external work Low levels of movement artefact Increased ease of blood sampling and BP monitoring Lower levels of apprehension Lower cost and space requirement
Negative Poor quantification of external work Increased apprehension Significant movement artefact Reduced ease of blood sampling and BP monitoring Greater space and cost requirement	*Negative* Unfamiliar mode Smaller muscle mass results in lower $\dot{V}O_2$max Lower limb fatigue limits performance Reduced control of work rate

altering speed (miles or kilometers·hour^{-1}), gradient (%), or both. For cycle ergometer tests, load is increased by altering resistance/power output (watts, W). To achieve optimal cardiorespiratory stress prior to other physiological limiters such as carbohydrate depletion, hyperthermia, and dehydration, protocols elicit cessation of exercise in less than 15 minutes. Very aggressive protocols (i.e., those that elicit maximal volitional exhaustion in less than 10 minutes) should not be used because acidosis and small-muscle fatigue may be the limiting factor rather than the cardiorespiratory system.

A detailed examination of protocols and their use is beyond the scope of this book; however, some basic protocols are commonly used in clinical and performance settings. In general, protocols are one of two types: step and ramp. Step protocols increase the intensity of exercise in a stepwise fashion (an instantaneous increase in workload) with large incremental steps in exercise load at set intervals (e.g., 50 W every three minutes). In contrast, ramp protocols employ a stepwise increase in which the steps are much smaller and shorter (e.g., 5 W every 15 s). A large number of protocols have been designed for use in a clinical setting (see figure 2.14). Among these examples of clinical step and ramp protocols in common clinical use

include the Bruce protocol and the Balke protocol, respectively (see figure 2.14). Each step and ramp protocol results in a specific cardiorespiratory response (see figure 2.15 on p. 33).

Step and ramp protocols have a number of limitations. Step protocols can be continuous or discontinuous (intermittent). Because discontinuous protocols alternate work and rest intervals, they lead to plateaus in heart rate, $\dot{V}O_2$, and blood pressure more readily than continuous protocols do. Significantly more time is required to complete discontinuous protocols, and although they may be advantageous for some patient populations (those with very low exercise capacity), continuous protocols are more widely used in clinical practice.

In comparison to ramp protocols, step protocols generally employ relatively aggressive (large) increases in exercise intensity that can lead to early test termination associated with noncardiorespiratory limitations, such as lower-limb (calf) pain. However, aggressive increases in workload limit the time required to attain maximum or target heart rate, which makes step protocols attractive in busy clinical practices.

Ramp protocols allow a closer interrogation of the chronotropic, blood pressure, and oxygen uptake responses to exercise. In particular, ramp protocols, unlike step

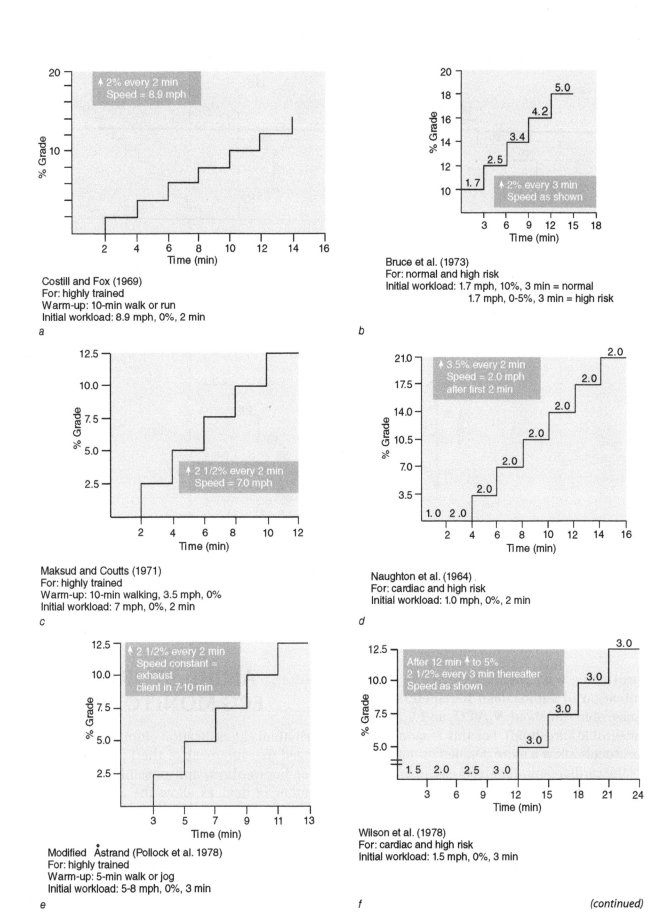

Costill and Fox (1969)
For: highly trained
Warm-up: 10-min walk or run
Initial workload: 8.9 mph, 0%, 2 min

a

Bruce et al. (1973)
For: normal and high risk
Initial workload: 1.7 mph, 10%, 3 min = normal
1.7 mph, 0-5%, 3 min = high risk

b

Maksud and Coutts (1971)
For: highly trained
Warm-up: 10-min walking, 3.5 mph, 0%
Initial workload: 7 mph, 0%, 2 min

c

Naughton et al. (1964)
For: cardiac and high risk
Initial workload: 1.0 mph, 0%, 2 min

d

Modified Åstrand (Pollock et al. 1978)
For: highly trained
Warm-up: 5-min walk or jog
Initial workload: 5-8 mph, 0%, 3 min

e

Wilson et al. (1978)
For: cardiac and high risk
Initial workload: 1.5 mph, 0%, 3 min

f

(continued)

Figure 2.14 Examples of protocols in common clinical use.

a-i Reprinted, by permission, from V. Heyward, 2010, *Advanced fitness assessment and exercise prescription,* 6th ed. (Champaign, IL: Human Kinetics), 75.

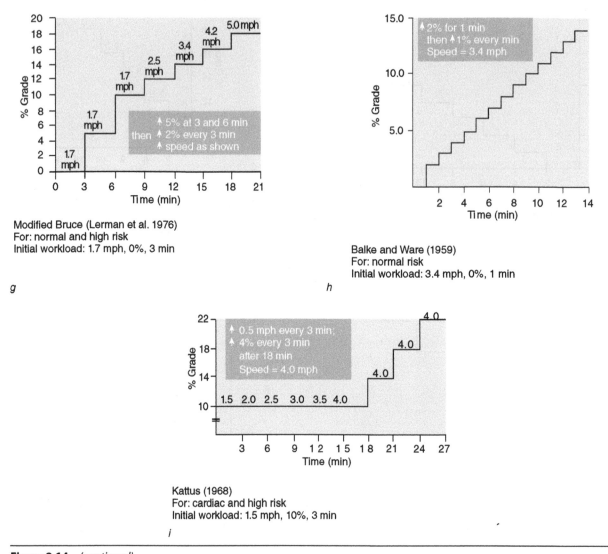

Modified Bruce (Lerman et al. 1976)
For: normal and high risk
Initial workload: 1.7 mph, 0%, 3 min

g

Balke and Ware (1959)
For: normal risk
Initial workload: 3.4 mph, 0%, 1 min

h

Kattus (1968)
For: cardiac and high risk
Initial workload: 1.5 mph, 10%, 3 min

i

Figure 2.14   *(continued)*

protocols, lend themselves to the determination of submaximal parameters (e.g., anaerobic threshold, V_E/VCO_2 and $V_E/\dot{V}O_2$ at anaerobic threshold). For this reason, ramp protocols allow a more detailed examination of the cardiorespiratory response to exercise. They are more time consuming, however, and require a greater level of operator expertise and experience.

The choice of exercise mode and protocol should be based on symptoms, the mode of exercise most likely to precipitate symptoms, and physical capacity.

AMBULATORY ECG MONITORS

Arrhythmia identification often requires prolonged measurement of the ECG waveform over hours, days, and possibly months. A number of devices allow the measurement and interpretation of the ECG over short periods of time (24 hours to 7 days). Short-term ECG monitors are termed ambulatory monitors or Holter monitors.

Ambulatory ECG monitors (Holter monitors) are portable ECG systems that record the elec-

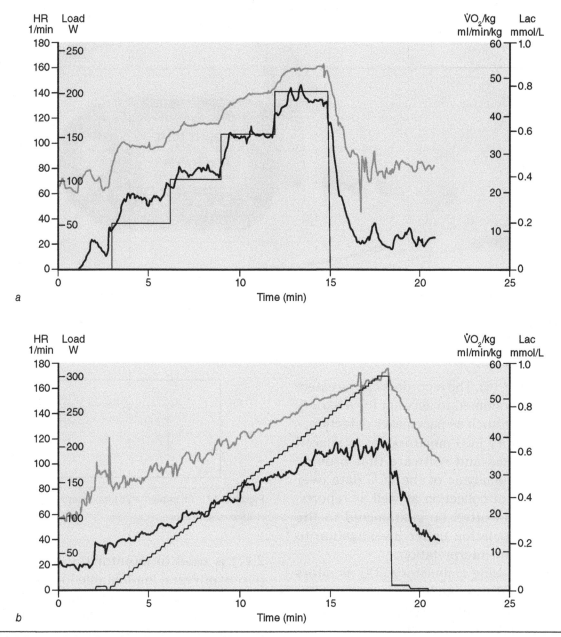

Figure 2.15 Integrated cardiopulmonary stress test graphs of *(a)* step and *(b)* ramp protocols demonstrating workload and related heart rate and oxygen consumption responses.

trical activity of the heart during normal daily life. Early versions of ambulatory monitors used magnetic tape to record the ECG, which made them relatively large, heavy, and slow to download data. Modern ambulatory monitors store data on hard drives and therefore are compact and lightweight and are downloaded rapidly using plug-and-play interfaces.

Most ambulatory systems collect information from three leads (see chapter 3) and have a screen display of the ECG waveform to ensure optimal lead placement. Ambulatory monitors have integrated event recorders allowing the patient to identify episodes during which they experience symptoms by simply pressing a button on the recorder

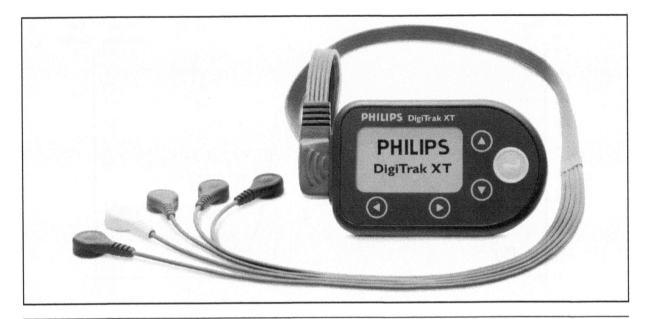

Figure 2.16   An ambulatory ECG monitor.

Photo courtesy of Philips.

(see figure 2.16). The recorders can be extensively programmed to account for a variety of scenarios such as pacemaker detection.

Ambulatory ECG monitors are interfaced with systems and software that provide automated analysis of the ECG data over the period of collection as well as reports. Reports can often be customized to the patient population under investigation to optimize data interpretation.

Patients using ambulatory ECG monitors are required to complete event diaries in which they document symptoms so these can be correlated with rhythm abnormalities. Event diaries differentiate benign arrhythmias (especially ectopics) in the absence of symptoms from those more sinister that are associated with symptoms.

IMPLANTABLE LOOP RECORDERS

A cardiac event such as an arrhythmia or syncope (fainting) that occurs only occasionally and sporadically is often difficult to monitor using an external ambulatory device. In this case an implantable loop recorder (see figure

Figure 2.17   An implantable loop recorder.

Photo courtesy of Philips.

2.17) is used to monitor the heart over a period of years. Implantable loop recorders are small devices usually implanted under the skin just below the left clavicle. The implantable recorder monitors the heart on a continuous loop, storing only abnormal rhythms that are detected automatically or, as available with some systems, when a handheld device is activated. Information from the device can be downloaded using telemetry so that it does not have to be explanted. Implantable loop recorders are interfaced with systems and software that provide automated analysis of the ECG data over the period of collection as well as reports. Reports can often be customized to the patient population under investigation to optimize data interpretation.

KEY POINTS

▶ Heart rate and rhythm can be determined very simply from auscultation (listening) of the heart with a stethoscope or palpation of a peripheral artery (i.e., carotid, radial, or tibial).

▶ The electrocardiograph automatically interprets waveforms providing an instant assessment of the ECG using algorithms that are specific to each manufacturer. Users should be careful about relying solely on the automated interpretations of commercial ECGs; a careful examination of the ECG should also be used.

▶ Preparation of the skin to reduce electrical impedance and accurate lead placement are crucial in obtaining a high-quality ECG trace.

▶ In general, during exercise the limb leads are moved from the ankles and feet to the anterior (front) surface of the body. The leads can also be placed on the posterior (back) surface of the body to improve the signal, particularly in obese people.

▶ Using tension loops and wireless client modules and securing leads together can significantly reduce artefact during exercise and improve the quality of the ECG trace.

▶ Ambulatory monitors and implantable loop recorders monitor the heart over extended periods of time (days to months).

THE ECG

Part II examines the ECG in detail at rest and during exercise. Chapter 3 focuses on obtaining a 12-lead ECG. Chapter 4 provides an examination of normal and abnormal ECG findings with a detailed examination of cardiac abnormalities. Each abnormality is accompanied by a sample "real world" ECG trace. Chapter 5 focuses on the ECG during exercise. Exercise creates a unique physiologic situation that results in a stimulating challenge to the cardiovascular system. As a result, the ECG during exercise has become a tool in the identification of a variety of pathologies, including coronary artery disease, exercise-induced arrhythmias, and assistance in the differentiation of pathologic and physiologic enlargement of the left ventricle. An increasing understanding of the importance of the exercise ECG in identifying and differentiating various diseases has resulted in an increased use of exercise as a diagnostic tool in cardiology.

3

Normal ECG at Rest

The ability to interpret the normal ECG at rest is fundamental to the interpretation of the abnormal resting and exercise ECG. A successful interpretation requires a high-quality ECG output, which results from an understanding of paper speed, calibration, skin preparation, and lead placement. Once a high-quality ECG output has been acquired, a standardized approach is used to interpret the ECG and report the findings. Measuring the key events on the ECG provides an understanding of the electrical and associated structural features of the heart. Identification of heart rate and rhythm is followed by the calculation of the mean QRS axis. Having established axis, the duration, amplitude, and nomenclature (shape) of the electrical events of the cardiac cycle are assessed. Normal limits exist for the following events: P wave, PR interval, QRS complex, ST segment, and QT interval. This chapter details the assessment of the normal 12-lead ECG at rest and highlights some of the common mistakes made in acquiring and interpreting a resting ECG.

READING THE ECG

The electrocardiograph records the electrical events of the cardiac cycle and presents them on an electrocardiogram (ECG). The ECG is a two-dimensional representation of a three-dimensional object, the heart. In this section, paper speed and calibration are discussed, along with how the leads work to create an ECG.

Paper Speed and Calibration

The ECG is printed onto paper that is divided into a grid of 1 mm square (1 mm^2) boxes. For ease of measurement and interpretation, the boxes are divided into continuous 5 mm^2 boxes bordered by heavier horizontal and vertical lines (see figure 3.1). The paper usually passes through the electrocardiograph at a speed of 25 mm·s^{-1}. The horizontal displacement of the ECG represents time and, based on a paper speed of 25 mm·s^{-1}, 1 mm is 0.04 s. The vertical displacement of the ECG represents the strength of the electrical signal measured in millivolts (mV). The electrocardiogram is usually standardized as 1 mm representing 0.1 mV (the signal can be set to half or double the normal sensitivity when the signal is very high or very low, respectively). Accordingly, the larger 5 mm^2 boxes represent 0.2 s horizontally and 0.5 mV vertically (see figure 3.1).

Ensuring correct standardization is crucial for the accurate interpretation of the ECG. To that end, the electrocardiograph produces a calibration signal (standardization mark) at the start of every trace. The standardization mark is a square wave 5 mm wide and 10 mm high representing 0.2 s duration (1 mm = 0.04 s) and 1 mV amplitude (1 mm = 0.1 mV; see figure 3.1). The first step in assessing an ECG is checking the presence, size, and accuracy of the standardization mark.

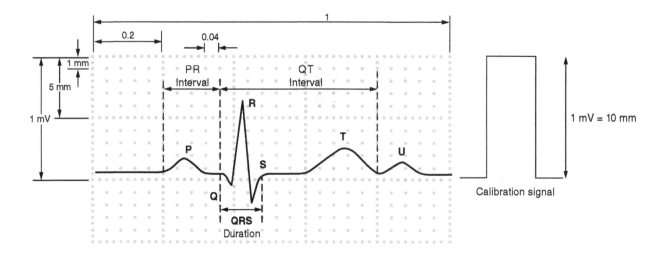

Figure 3.1 ECG paper demonstrating grid lines and standardization mark.

Time and voltage

Vertical axis:
- 1 small square = 1 mm (0.1 mV)
- 1 large square = 5 mm (0.5 mV)
- 2 large squares = 10 mm (1.0 mV)

Horizontal axis:
- 1 small square = 0.4 s
- 1 large square = 0.2 s
- 5 large squares = 1.0 s

Limb and Chest Leads

The ECG is a two-dimensional representation of a three-dimensional object, the heart. To overcome this fundamentally limiting characteristic, a number of electrodes (leads) are placed in various positions on the body to create a three-dimensional model of the heart. The position of these leads allows the observer to view the heart from the top (the base) to the bottom (the apex) and from the front (anterior) to the back (posterior), enabling an interrogation of the electrical properties of all areas of the heart. Ten electrodes are assigned as limb leads (4 electrodes) and chest (precordial) leads (6 electrodes) in a routine standard 12-lead ECG. Each ECG lead is at a unique position and distance from the heart, the leads demonstrate differences in orientation and voltage and therefore, shape.

Limb Leads

As discussed in chapter 2, in the resting 12-lead ECG the limb leads are positioned on the right and left ankle and right and left wrist and are commonly assigned the following labels: right leg (RL), left leg (LL), right arm (RA), and left arm (LA). The right foot acts as the neutral lead and is usually assigned the label N (see figure 3.2). The remaining three leads act in two electrically different ways to propagate six distinct leads on the

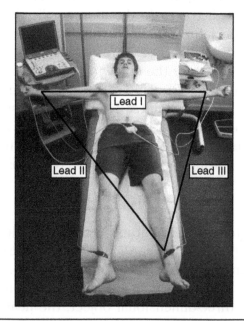

Figure 3.2 The relationship between Einthoven's triangle and limb lead placement.

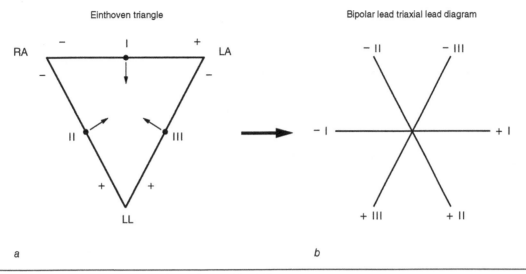

Figure 3.3   Einthoven's triangle and the derived triaxial lead diagram.

ECG. The two subgroups of limb lead are termed unipolar (augmented) and bipolar. Each lead acts in both unipolar and bipolar capacities.

The bipolar leads are labeled lead I, lead II, and lead III. Each lead measures the electrical difference between two leads in the following format:

$$\text{Lead I} = \text{RA (–ve)} \rightarrow \text{LA (+ve)}$$

$$\text{Lead II} = \text{RA (–ve)} \rightarrow \text{LL (+ve)}$$

$$\text{Lead III} = \text{LA (–ve)} \rightarrow \text{LL (+ve)}$$

Based on these formulas, the relationship that exists among the bipolar leads can be expressed as follows: lead I + lead III = lead II.

Leads I, II, and III are represented schematically as Einthoven's triangle (named after the Dutch physician who invented the ECG). The triangle can be redrawn so that leads I, II, and III intersect at a common central point resulting in a triaxial diagram (see figure 3.3).

The unipolar augmented (a) vector (V) leads are labeled aV_R, aV_L, and aV_F. The unipolar leads measure the electrical signal at one location relative to zero potential (i.e., the signal received directly by the electrode). The augmentation of the signal is associated

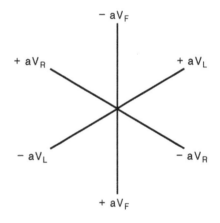

Figure 3.4   Unipolar triaxial diagram.

with a 50% amplification of the signal strength to make the complexes easier to read and interpret. As with Einthoven's triangle, a triaxial diagram is used to represent the spatial orientation of the augmented limb leads (see figure 3.4). The relationship among the augmented leads is as follows: $aV_R + aV_L + aV_F = 0$. This relationship means that the sum of the voltages for each complex (e.g., P wave, QRS, or T wave) is zero.

Each limb lead has a specific polarity and spatial orientation. Accordingly, combining the preceding two triaxial representations provides a hexaxial diagram of the limb leads (see figure 3.5).

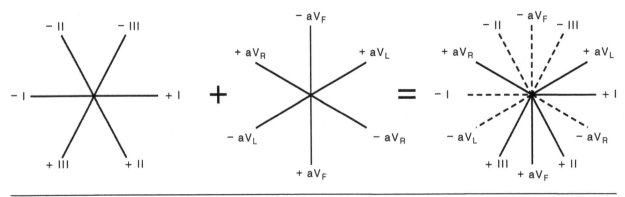

Figure 3.5 The bipolar and unipolar triaxial diagrams are used to derive the hexaxial lead diagram. The negative pole of each lead is represented by a dashed line.

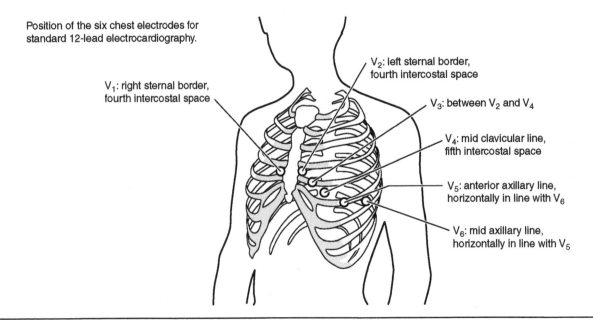

Position of the six chest electrodes for standard 12-lead electrocardiography.

V_1: right sternal border, fourth intercostal space

V_2: left sternal border, fourth intercostal space

V_3: between V_2 and V_4

V_4: mid clavicular line, fifth intercostal space

V_5: anterior axillary line, horizontally in line with V_6

V_6: mid axillary line, horizontally in line with V_5

Figure 3.6 Chest lead positions.

Chest Leads

The chest leads are labeled V_1 to V_6 and are positioned in the following manner (see figure 3.6):

- V_1—Fourth intercostal space, right sternal border
- V_2—Fourth intercostal space, left sternal border
- V_3—Midway between V_2 and V_4
- V_4—Fifth intercostal space, midclavicular line
- V_5—Midway between V_4 and V_6
- V_6—Midaxillary line at the same level as V_4

The schematics for the limb leads and chest leads provide a frontal and horizontal plane, respectively. The limb leads record voltages directed upward and downwards and to the right and left of the heart. In contrast, the chest leads record voltages directed anterior and posterior and to the right and left (see figure 3.7). Each lead can be seen as a camera obtaining a picture of the heart from 12 directions. Accordingly, combinations of leads can be used to interrogate various aspects of the heart (see table 3.1).

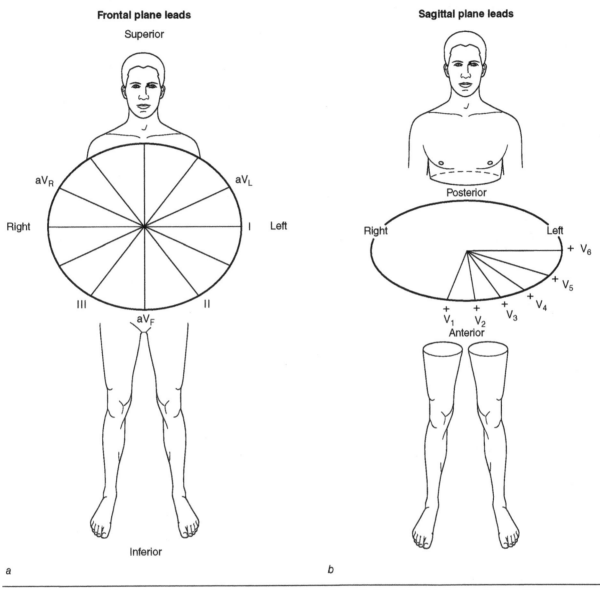

Frontal plane leads

Superior

aV_R

aV_L

Right

I Left

III

II

aV_F

Inferior

a

Sagittal plane leads

Posterior

Right

Left

+ V_6

+ V_5

+ V_4

+ V_3

+ V_1 + V_2

Anterior

b

Figure 3.7 The spatial relationship of the limb and chest leads as shown in the *(a)* frontal plane and *(b)* sagittal plane.

Table 3.1 Anatomical Relations of Leads in a Standard 12-Lead ECG

Leads	Anatomical relation
II, III, aV_F	Inferior surface of the heart (apex)
V_1 to V_4	Anterior surface of the heart (right ventricle and interventricular septum)
I, aV_L, V_5, and V_6	Lateral surface of the heart (left ventricle)
V_1 and aV_R	Right atrium and cavity of left ventricle

INTERPRETING THE ECG

The process of depolarization and repolarization results in the characteristic ECG waveform. The key electrical events of the cardiac cycle result in the P-QRS-T-U wave and are a measure of the electrical properties of the heart reflecting the conduction pathway and velocity across the myocardium. Measuring the key events on the ECG allows an interrogation of the electrical and associated structural features of the heart. Normal limits exist for the following events: P wave, PR interval, QRS complex, ST segment, and QT interval. Furthermore, the morphology (shape) of each of the key events on the ECG is used in the interpretation of the ECG. Prior to measuring each ECG complex, observers conventionally measure the heart rate, observe the rhythm, and calculate the electrical axis of the heart.

Heart Rate

Having noted the standardization signal, the ECG observer establishes the rate (the number of cardiac cycles over a defined period of time, usually a minute). A normal heart rate is 60 to 100 beats·min^{-1}. Slower rates are termed bradycardia (<60 beats·min^{-1}, although this may be normal in athletes), and faster rates are termed tachycardia (>100 beats·min^{-1}). Most electrocardiographs automatically measure heart rate; however, this ability may be compromised in the presence of certain arrhythmias. Furthermore, atrial rate may be different from the ventricular rate in people with certain conditions (e.g., complete heart

block). Therefore, an ability to manually calculate the rate is crucial. Following are three simple methods for measuring heart rate.

1. A very quick, but not the most accurate, method of measuring heart rate is to find a QRS complex that occurs on the heavy line of a 5 mm box and simply count across consecutive large (5 mm) boxes in the following sequence until you reach the next QRS complex: 300, 150, 100, 75, 60, 50, 40. Using the number assigned to the heavy line occurring immediately prior to the next QRS complex is the heart rate (see figure 3.8a).

2. Count the number of large horizontal boxes (5 mm, 0.2 s) between two successive R waves (or P waves, if atrial rate is of interest). To improve this calculation, select an R wave that falls on the solid line of the large box. Having established the number of large boxes, divide a constant (300) by this number.

- Heart rate (beats·min^{-1}) = 300 / number of large boxes (see figure 3.8b)

- Alternatively, if a more accurate measure of rate is required divide a constant (1,500) by the number of small boxes (1 mm, 0.04 s).

- Heart rate (beats·min^{-1}) = 1,500 / number of small boxes (see figure 3.8c)

3. When the heart rate is irregular, the preceding methods are difficult to employ. In these cases, calculating an average rate is the most informative way to present heart rate. Simply count the number of cardiac cycles (R waves, or P waves if atrial rate is different) in

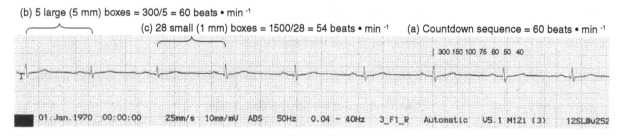

(b) 5 large (5 mm) boxes = 300/5 = 60 beats • min^{-1}

(c) 28 small (1 mm) boxes = 1500/28 = 54 beats • min^{-1} (a) Countdown sequence = 60 beats • min^{-1}

| 300 150 100 75 60 50 40

01.Jan.1970 00:00:00   25mm/s 10mm/mU ADS 50Hz 0.04 – 40Hz 3_F1_R Automatic U5.1 M121 (3)   12SL8u252

Figure 3.8   Calculating heart rate using (a) a simple countdown sequence, (b) 300 / the number of large boxes, and (c) 1,500 / the number of small boxes.

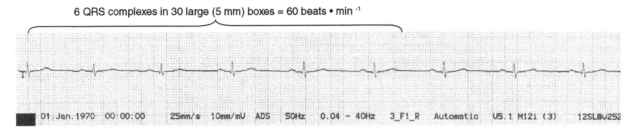

6 QRS complexes in 30 large (5 mm) boxes = 60 beats • min⁻¹

01.Jan.1970 00:00:00   25mm/s 10mm/mV ADS 50Hz 0.04 - 40Hz 3_F1_R Automatic V5.1 M12i (3)   12SL8v252

Figure 3.9 Calculating heart rate using an average over a 6 s period.

6 s (30 large boxes) and multiply by 10. This task is assisted by the presence of 3 s (15 large boxes) markers found at the top of ECG paper (see figure 3.9).

Rhythm Recognition

Identifying the rhythm is an important first step in the interpretation of the ECG. The vast majority of rhythm disturbances can be determined by answering the following simple questions:

- What is the atrial rhythm?
- What is the ventricular rhythm?
- Is the AV conduction normal?
- Are there any unusual complexes?
- Is the rhythm dangerous?

Chapter 4 examines rhythm disturbances in more detail. Following are basic things to consider when answering the preceding questions.

- Identifying the origin of atrial depolarization is fundamental in establishing atrial rhythm. Establishing whether the atria are activated from the SA node, atria, or AV junction is important in the identification of arrhythmia. In the normal heart, P waves should be clearly identifiable, occurring prior to every QRS complex. A sinus origin is assumed if the P waves are positive in lead II, aV_F, and V_1, and if the rate is normal (60-100 beats·min⁻¹), sinus rhythm is identified.

- Identifying the origin of ventricular depolarization is fundamental in establishing ventricular rhythm. Establishing whether the ventricles are activated from a supraventricular source (sinoatrial, atrial, or atrioventricular) or a ventricular source (including Purkinje's system) is important in the identification of arrhythmia. In the normal heart the QRS complex is preceded by a P wave at a fixed, normal PR interval (0.12-0.2 s).

- AV conduction is assessed by examining the relationship between the P waves and the QRS complexes. In the normal heart, sinus rhythm is present if the QRS complexes are always preceded by P waves with a fixed, normal PR interval (0.12–0.2 s).

- P waves, QRS complexes, and T waves have the same normal nomenclature (size and shape) in a single lead.

- Sinus rhythm of a normal resting rate (50-100 beats·min⁻¹, although this may be lower in young, fit people) is not considered dangerous.

Electrical Axis and Its Determination

Understanding three fundamental ECG rules allows a detailed interrogation of cardiac electrical activity:

1. A positive deflection is observed when the wave of depolarization moves toward the positive pole of a lead.

2. A negative deflection is observed when the wave of depolarization moves away from the positive pole of a lead.

3. A biphasic (upward and downward deflections of the same magnitude) wave is observed when the wave of depolarization is perpendicular to the lead.

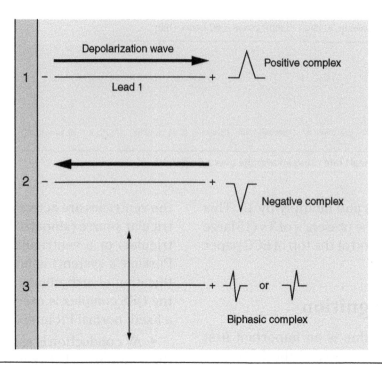

Figure 3.10 The ECG shows a positive complex in any lead where the wave of depolarization spreads towards the positive pole of the lead (1) and a negative complex where the wave of depolarization spreads away from the positive pole of the lead (2). When the spread of the depolarization wave is perpendicular to the lead, a biphasic (combined positive and negative deflection) wave is generated (3).

The same rules hold for repolarization (see figure 3.10).

The initiation of the action potential occurs at the SA node and spreads across the atria (assisted by the atrial internodal bands) in a rapid wavelike depolarization from the basal right atrium distally to the left atrium toward the AV node. The depolarization wave at a single time point has a direction and magnitude. Therefore, the depolarization of myocardial cells can be represented as electrical vectors. The sum of these vectors is represented by a single resultant vector demonstrating the direction and magnitude of depolarization and is represented by an arrow varying in size to reflect the magnitude of depolarization (see figure 3.11). The larger the muscle mass, the larger the magnitude of the resultant vector, and therefore, the greater the deflection observed on the ECG. Atrial depolarization is represented by a small, short-duration deflection on the ECG, the P wave (see figure 3.12 on p. 48). Leads with an anterior, lateral, or inferior positive pole show a predominantly positive P-wave deflection (i.e., leads I, II, III,

aV_F, aV_L, and V_2 to V_6). In contrast, leads with a posterior or superior positive pole show a predominantly negative P-wave deflection (i.e., leads aV_R and V_1).

Although the depolarization of the ventricles is more complex than that of the atria; the same basic rules apply. The conduction of the action potential through the specialized tissues of the AV junction, bundle branches, and Purkinje's system results in a rapid depolarization of the ventricles through a recognized sequence of events. The first rapid (<0.04 s) phase of depolarization occurs in the interventricular septum. Because the left edge of the septum is first to depolarize because it emanates from a branch of the left bundle of His, the wave of depolarization spreads from left to right with a resultant electrical vector in the same direction. The magnitude of this vector is small given the small muscle mass of the interventricular septum and is represented by a small deflection on the ECG (see figures 3.11 and 3.12). Depending on the position of the positive pole of the lead, septal depolarization is represented as a Q wave (often noted as *q wave* because of its

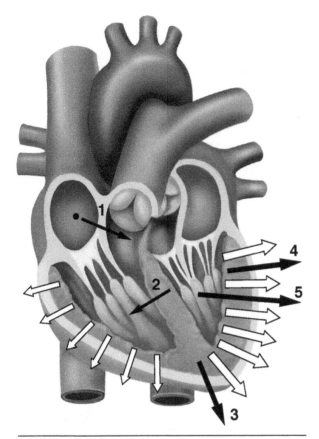

Figure 3.11 Atrial depolarization represented by a resultant vector directed inferiorly and leftward (1). The ventricular myocardium depolarizes from endocardium to pericardium with a magnitude of electrical charge related to the myocardial mass leading to the development of electrical vectors (open arrows; note the smaller vectors for the right ventricle associated with a smaller muscle mass). Ventricular depolarization commences with septum (2) followed by apical and early left ventricular depolarization (3) and ending with late ventricular depolarization of the basal ventricular myocardium (4). The sum of these individual vectors creates the resultant depolarization vector (5) measured by the electrocardiogram.

low magnitude [<0.4 mV]) or as an R wave (again, often noted as *r wave* because of its low magnitude).

Depolarization of the septum is rapidly followed by the apical and early left ventricular depolarization. Because the ventricular myocardium depolarizes from the endocardium to the epicardium, apical and early left ventricular depolarization give rise to a resultant electrical vector that is directed inferiorly and leftward (see figure 3.11). Late ventricular depolarization involves the lateral wall of the left ventricle and the posterior wall

of the right ventricle propagating a resultant vector directed superiorly and leftward (see figure 3.11). The significantly larger muscle mass of the left ventricle results in a large-magnitude resultant vector directed inferiorly and leftward and represented by a large QRS complex on the ECG (see figure 3.12).

A short period of electrical inactivity follows ventricular depolarization, during which time the ECG returns to the isoelectric line (the ST segment). Following this, the repolarization wave of the ventricles spreads from the endocardium to the epicardium at a slower rate than that observed in ventricular depolarization leading to a longer-duration, lower-magnitude wave on the ECG (T wave; see figure 3.12).

The resultant electrical vectors of atrial and ventricular depolarization and ventricular repolarization can be used to describe the mean electrical axis. The mean electrical axis describes the net direction of electrical activity in the frontal plane toward which the P wave, QRS complex, and T wave are pointing. Calculating the mean electrical axis is valuable in the initial evaluation of the anatomic position of the heart and the direction of depolarization. In general, the mean QRS axis is of primary interest and is the focus of attention. It is calculated using the limb leads (frontal plane). The hexaxial lead diagram, adapted to include angular designations, is used in the calculation of the mean QRS axis (see figure 3.13 on p. 49).

A normal mean QRS axis is between –30° and +100°. Left axis deviation is considered to occur between –30° and –90°. Right axis deviation is between +100° and 180°, and extreme axis deviation is between –90° and 180° (see figure 3.14 on p. 49).

Calculating the Mean QRS Axis

Although the atrial electrical axis is rarely measured, the vast majority of automated electrocardiographs display the mean P-wave axis. The mean T-wave axis is significant in certain circumstances, such as when locating specific areas of ischemic or infarcted myocardium.

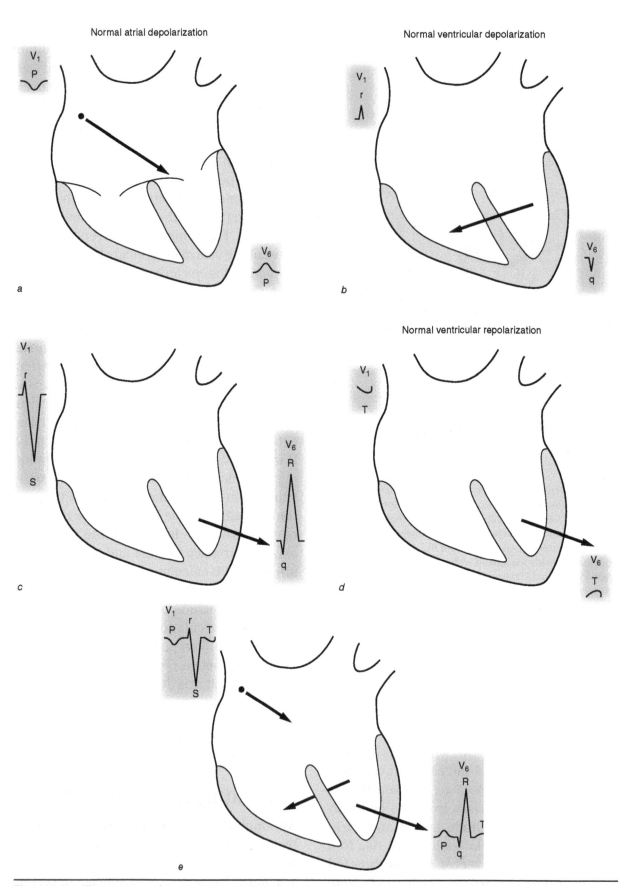

Figure 3.12 The sequence of depolarization and repolarization of the heart demonstrating the resultant electrical vectors and the corresponding ECG trace: *(a)* atrial depolarization; *(b)* depolarization of the septum; *(c)* resultant vector of apical and early left ventricular and late ventricular depolarization; *(d)* ventricular repolarization; *(e)* summary of cardiac electrical activity.

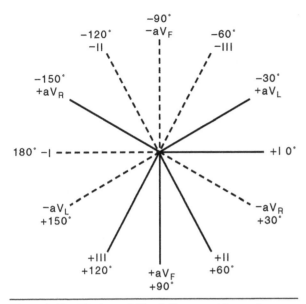

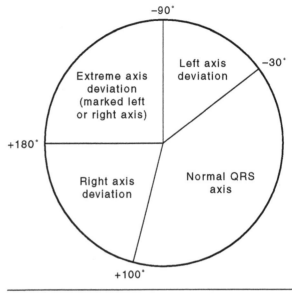

Figure 3.13 The hexaxial lead diagram with angular assignment combining the triaxial diagrams for all 6 limb leads. Each lead has a +ve and –ve (dashed line) pole that is assigned an angular value.

Figure 3.14 Schematic of hexaxial lead diagram demonstrating normal QRS axis, right and left axis deviation, and extreme axis deviation.

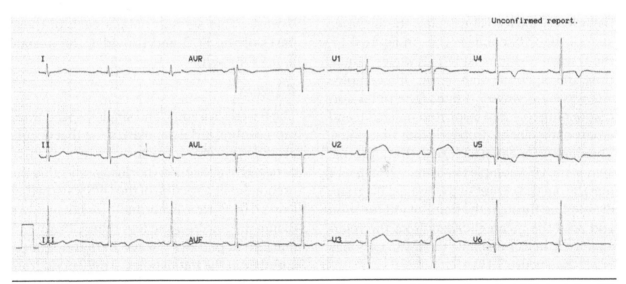

Figure 3.15 A normal mean QRS axis calculated using method 1. Note the two tall R waves of equal height in leads II and III. The mean QRS axis lies midway between lead II and lead III, at approximately 90°.

The mean QRS axis is the most commonly measured parameter, and although a precise calculation is complicated, the following three simple calculation methods are useful.

1. The mean QRS axis points midway between any two leads that show tall R waves of equal height (see figure 3.15).

2. The mean QRS axis lies perpendicular to any lead showing a biphasic QRS wave.

Because a positive deflection is observed when the depolarization wave of the ventricles is moving toward the positive pole of a lead, the mean QRS axis will be pointing toward leads with a positive QRS deflection (see figure 3.16 on p. 50).

3. The mean QRS axis can be calculated accurately by plotting vectors on a bipolar lead (I, II, III) triaxial reference grid (a rearranged Einthoven's triangle; see figure 3.17).

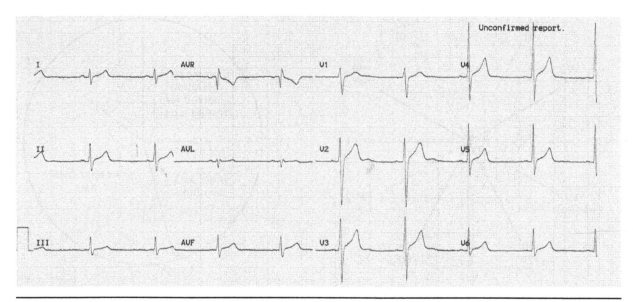

Figure 3.16 A normal mean QRS axis calculated using method 2. Note the biphasic wave in lead I. Accordingly, the mean QRS axis lies perpendicular to lead I and directed toward the positive poles of leads II and III (leads II and III demonstrate positive QRS complexes indicating the direction of the QRS resultant vector toward these leads). The mean QRS axis is therefore approximately 90°.

Using lead I and lead III, calculate the algebraic sum of the QRS complex (i.e., in figure 3.17, the Q wave = –1 mm, and the R wave = 5 mm; thus, the algebraic sum of the QRS complex = Q wave + R wave = 4 mm). The point corresponding to 4 mm is located on lead I, and a perpendicular is dropped from that point. The same is repeated for lead III (the algebraic sum of the QRS complex = –3 mm + 17 mm = 14 mm). From the center of the triaxial diagram, draw a line through the point of intersection and read the mean QRS axis from the reference grid. In this example the mean QRS axis is 77° (see figure 3.17).

Methods 1 and 2 are estimates of the mean QRS axis, and R waves (or S waves) of approximately similar heights or QRS complexes that are close to biphasic can be employed in the calculation. An error of 10 to 15 is clinically acceptable; however, if accurate measurement of the mean QRS axis is necessary, method 3 should be employed.

Measurement and Normal Limits

The ECG can be broken down into the P wave, PR interval, QRS wave, QRS complex, QT interval, ST segment, the T wave, and the U wave. This section details each of these parts of the ECG and provides the normal limits for each.

P Wave

The P wave is a small and single-peaked wave representing atrial depolarization that occurs prior to the QRS wave and can be a positive or negative deflection (normally reflecting the predominant direction of the QRS wave in that lead). The P wave is normally less than 0.12 s (3 mm) in duration and less than 0.25 mV (2.5 mm) in amplitude. P-wave duration is measured from the initial point of deflection from the isoelectric line to the return of the wave to the isoelectric line. The amplitude of the P wave is measured from the isoelectric line to the peak of the P wave. The mean P-wave axis is oriented most directly toward leads II and aV_F in the frontal plane and V_1 in the sagittal plane. Because the largest P waves are seen in these leads, these leads should be used to examine P-wave rhythm and nomenclature (shape; see figure 3.1 on p. 40).

PR Interval

The PR interval is the time between the initiation of atrial depolarization and the start of

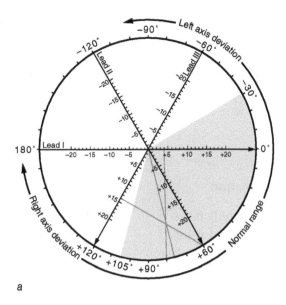

a

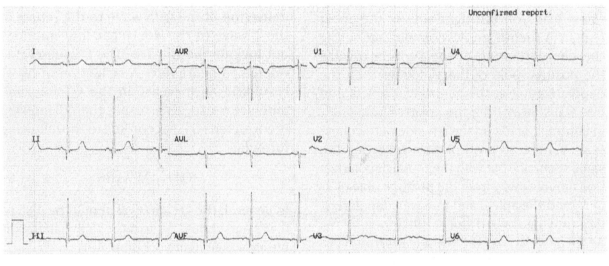

b

Figure 3.17 Calculation of the mean QRS axis using method 3. In this example the algebraic sum of the QRS in lead I = 4 mm and in lead III = +14 mm. Lines are drawn, and the mean QRS axis = 77°.

ventricular depolarization. The PR interval reflects the time the action potential takes to spread from the SA node across the atria and through the AV junction (AV node and bundle of His). The PR interval is measured from the initial point of deflection from the isoelectric line of the P wave to the initial deflection of the QRS wave from the isoelectric line. The PR interval is normally between 0.12 s and 0.2 s (3 to 5 mm) in duration. The length of the PR interval may vary in different ECG leads, and the shortest should be noted (see figure 3.1 on p. 40).

QRS Wave

The QRS wave represents the depolarization of the ventricles. The QRS duration is measured from the first deflection of the PR segment from the isoelectric line (start of the QRS complex) to the return of the QRS complex to the isoelectric line. In the normal heart the QRS complex is less than 0.10 s in duration. The nomenclature (shape) of the QRS complex depends on the lead that is used to observe the complex. As detailed previously, the nomenclature of the QRS wave depends on the position of the positive pole of the lead and the direction of the depolarization of the ventricles. Although the resultant QRS vector has a single magnitude and direction, the QRS complex reflects the sequence of ventricular depolarization:

1. Septal
2. Apical and early left ventricular
3. Late ventricular depolarization

Accordingly, the QRS complex will have a different nomenclature in each lead related to the position of the positive pole of the lead.

QRS Complex in the Chest Leads

Chest leads on the right side of the chest (e.g., V_1) record an RS complex reflecting the depolarization of the septum toward the positive pole of the V_1 followed by the depolarization of the ventricular apical and free walls away from the positive pole of V_1. In contrast, leads on the left side of the chest (e.g., V_6) record a qR pattern reflecting the same sequence but with the septal depolarization moving away from the positive pole of V_5 or V_6 and the apical and free wall depolarization moving toward the positive pole of V_6. As we move across the chest leads from V_1 to V_6, the R wave becomes relatively larger and the S wave becomes relatively smaller. This is termed R-wave progression, and the peak R-wave amplitude normally occurs in leads V_4 to V_5 (a transition zone can be identified, usually around V_3 to V_4 where the R:S ratio becomes 1).

QRS Complex in the Limb Leads

As with the chest leads, each limb lead has a unique QRS nomenclature associated with the position of the positive pole of the lead. The positive pole of lead aV_R is superior and rightward, leading to a predominantly negative QRS complex (commonly observed with a negative T wave). The QRS complex in the remaining five leads is somewhat more complex because of the considerable normal variance observed in these leads. The pattern observed is closely associated with the mean QRS axis (described earlier). Although it is not possible to define normal QRS patterns, in general, leads II, III, and aV_F record a predominantly positive QRS complex, whereas leads I and aV_L record positive, biphasic, or very small-magnitude QRS complexes in the normal heart.

QT Interval

The QT interval is the time between the onset of ventricular depolarization and the end of the repolarization of the ventricles. The QT interval is measured from the initial point of deflection of the QRS wave to the return of the T wave to the isoelectric line and does not include the U wave. The QT interval is related to heart rate. A reduction in heart results in an increase in the QT interval, and vice versa. As a result, the QT interval is corrected (QT_c) according to the following formula:

$$QT_c = QT / \sqrt{RR}$$

In general, the QT interval should be 0.35 to 0.45 s and should be no more than half the RR interval (see figure 3.1 on p. 40).

ST Segment

The ST segment is the time between ventricular depolarization and repolarization. The ST segment is measured from the terminal portion of the QRS wave to the start of the T wave. The junction between the terminal portion of the QRS wave and the start of the ST segment is called the J point. In the normal heart the ST segment resides on the isoelectric line (see figure 3.1 on p. 40).

T Wave

The T wave represents ventricular repolarization and is measured from the first deflection of the ST segment from the isoelectric line to the return of the T wave to the isoelectric line. In general, the T wave is asymmetrical: the primary slope is less steep than the secondary slope. The orientation of the T wave normally reflects the orientation of the QRS complex. Although there is no definitive measure of T-wave amplitude, it typically corresponds to the height of the preceding R wave (i.e., the bigger the R wave, the bigger the T wave) and is considered normal when it is more than 1/8 but less than 2/3 the amplitude of the preceding R wave (see figure 3.1 on p. 40).

U Wave

The U wave is a small deflection that follows the T wave and is generally deflected in the same direction as the preceding T wave. The origin of the U wave is unclear; however, it has been suggested to represent the repolarization of the midmyocardial cells (those cells between the endocardium and epicardium) and the His–Purkinje system. The U wave is most commonly observed in V_2 to V_3; however, it is often not present on the ECG. Prominent U waves may be found in athletes and are associated with hypokalemia, hypercalcemia, and hyperthyroidism (see figure 3.1).

Table 3.2 provides a summary of the normal limits of the resting 12-lead ECG.

COMMON MISTAKES IN PREPARING AND PLACING LEADS (TROUBLESHOOTING)

Leads must be attached correctly to ensure correct positioning and optimal signals.

Skin Preparation

Artefact, or noise, on the ECG can make rhythm recognition, measurement, and interpretation very difficult. There are a number of ways to improve and optimize the signal received by the electrode. Preparation of the skin is fundamental to reducing electrical impedance and improving the signal intensity. The skin should be prepared using alcohol to remove surface grease followed by an abrasive pad to remove dead skin. The skin should be allowed to dry fully prior to the placement of the electrode. Shaving may be necessary to remove excessive hair (this is particularly important for exercise ECG).

ECG Electrodes

The quality of ECG electrodes can have a significant impact on the quality of trace obtained. Ensuring that the electrodes have

Table 3.2 Summary of the Normal Limits of the Resting 12-Lead ECG

	QRS axis	P wave	PR interval	QRS	QT interval	ST segment	T wave	U wave
Normal limits	−30° to +100°	<0.12 s (3 mm) and <0.25 mV (2.5 mm) (best measured in V_1, II, or aV_F)	0.12 s and 0.2 s (3 to 5 mm)	<0.1 s (variable shape)	0.35-0.45 s (no more than half the RR interval) ($QT_c = QT / \sqrt{RR}$	J point on the isoelectric line	Normally same orientation as QRS complex and >⅛ but <⅔ the amplitude of the preceding R wave	Small deflection normally in the same direction as the preceding T wave

not passed their manufacturer-recommended expiration date is important; however, a visual inspection of the electrodes to make sure that the gel, where present, has not dried out is the most effective quality check. Defective cables may also produce sharp fluctuations or an isoelectric waveform. If disturbances can be produced by gently pulling on the cable, the cable should be changed.

Muscle and Lead Movement

Occasionally, electrode movement or muscular activity can create artefact. For this reason, the patient must remain absolutely still while obtaining the ECG. Placing a tension loop in the lead may help with artefact if movement is uncontrollable (this is particularly true during exercise). Excessive movement of the leads across one another may also lead to artefact. Positioning and fixing the leads to minimize movement will reduce noise. The use of tension loops can significantly reduce lead movement.

Other electrical equipment in the vicinity of the electrocardiograph may cause interference leading to artefact on the ECG. Electrical filters (common in modern electrocardiographs) reduce this potential. Nevertheless, minimizing the number of electrical devices near the electrocardiograph and ensuring the best possible electrode contact is good practice.

Lead Placement

Incorrect lead placement is a common mistake. Practitioners can use a number of simple checks to confirm correct lead placement. The most obvious is to check the leads on the patient. Doing so often reveals mistakes. An examination of the limb leads on the ECG trace can reveal erroneously placed leads. Remember the following relationships: $aV_R + aV_L + aV_F = 0$ and lead I + lead III = lead II. If the sum of the augmented leads does not equal zero, or if the sum of leads I and III do not equal lead II, the leads are placed incorrectly.

KEY POINTS

▶ Always check the standardization mark prior to collecting and interpreting an ECG. Normally, the horizontal scale (duration) is 1 mm = 0.04 s and the vertical scale (amplitude) is 1 mm = 0.1 mV.

▶ To check limb lead placement, consider the following, which normally hold true:

$$aV_R + aV_L + aV_F = 0, \text{ and lead I + lead III = lead II.}$$

▶ Having established heart rate, assess the rhythm by asking the following questions: (1) What is the atrial rhythm? (2) What is the ventricular rhythm? (3) Is the AV conduction normal? (4) Are there any unusual complexes? (5) Is the rhythm dangerous?

▶ QRS axis is assessed using the QRS complexes in the limb leads. A normal mean QRS axis is between −30° and +100°

▶ ECG normal limits are as follows: P wave < 0.12 s and < 0.25 mV; PR interval = 0.12 to 0.2 s; QRS < 0.1 s; QT interval = 0.35 to 0.45 s (remember rate-corrected QT_c); ST segment presents with J point on the isoelectric line; T waves > 1/3 and < 2/3 of the preceding R wave and normally have the same orientation as QRS complex.

4

Abnormal ECG at Rest

This chapter is divided into five main sections: 1. bradyarrhythmias; 2. tachyarrhythmias; 3. atrial and ventricular enlargement; 4. myocardial ischemia and infarction; and 5. unusual ECG abnormalities. Each section provides examples of these conditions including key features along with common causes. Supporting ECGs are presented to illustrate their clinical presentation and assist in the recognition of the key features.

The first section examines arrhythmias associated with slow heart rates, termed bradyarrhythmias. These arrhythmias include axis deviation, sinus bradycardia, sinus arrhythmia, AV junctional rhythms, and blocks in electrical conduction associated with the SA node, AV junction, bundle branches, and fascicles. The second section examines arrhythmias associated with higher heart rates, termed tachyarrhythmias. This group of arrhythmias is subdivided into arrhythmias originating above the ventricles (supraventricular), arrhythmias originating in the ventricles (ventricular arrhythmias), elevated heart rate as a result of an arrhythmia originating in the ventricles (ventricular tachycardia), and elevated heart rate without an obvious origin (idiopathic ventricular tachyarrhythmia).

The third section examines ECG patterns commonly observed when the atria or ventricles are enlarged. The fourth section focuses on myocardial ischemia and the ECG presentations of subendocardial and transmural MI, NSTEMI, and STEMI. The text and accompanying ECG figures address the diagnosis of regions of infarction and related arteries. The final section examines a wide range of less common ECG abnormalities that are associated with specific disorders that do not fall into the categories previously outlined in the chapter.

Identifying the Cause of Abnormality

Answering the following questions will help identify and establish the cause of any abnormality:

- Has the standardization mark been checked?
- What is the heart rate?
- What is the electrical axis?
- What is the atrial rhythm?
- What is the ventricular rhythm?
- Is the AV conduction normal?
- Are there any unusual complexes?
- Is the rhythm dangerous?

BRADYARRHYTHMIAS

Bradyarrhythmias are unusual rhythms generally associated with low heart rates that fall under a number of categories, including low heart rate originating from the sinus node (sinus bradycardia); sinus rhythms with a marked variability in RR interval (sinus arrhythmia); rhythms originating in the AV junction (AV junctional rhythms); arrest of the sinus node depolarization leading to depolarization from an ectopic focus (sinus arrest and escape beats); blocking of the action potential through the AV junction (heart block including first, second [Mobitz types I and II], and third degree); and blocking of the action potential at the level of the bundle branches or fascicles (left and right bundle branch blocks [LBBB, RBBB, incomplete LBBB and RBBB, and fascicular blocks]). The following sections examine each of these conditions with supporting ECGs to illustrate their presentations.

Axis Deviation

A normal mean QRS axis is between –30° and +100°. Left axis deviation (LAD) is considered to occur between –30° and –90°. Right axis deviation (RAD) is between +100° and 180°, and extreme axis deviation is between –90° and 180° (see figure 3.14 on p. 49). The presence of axis deviation is not always indicative of an underlying cardiac pathology; however, it can support other ECG findings. Identification of RAD and LAD is relatively simple.

Right Axis Deviation (RAD)

In general, if leads II and III show tall R waves, with lead III greater than lead II, RAD is present. Furthermore, lead I will show an RS complex with the S wave deeper than the R wave is tall (see figure 4.1).

Following are common causes of RAD:

- Normal variant
- Right ventricular hypertrophy
- Myocardial infarction of the lateral wall of the left ventricle
- Left posterior hemiblock (rare)
- Chronic lung disease: emphysema and chronic bronchitis
- Acute pulmonary embolism

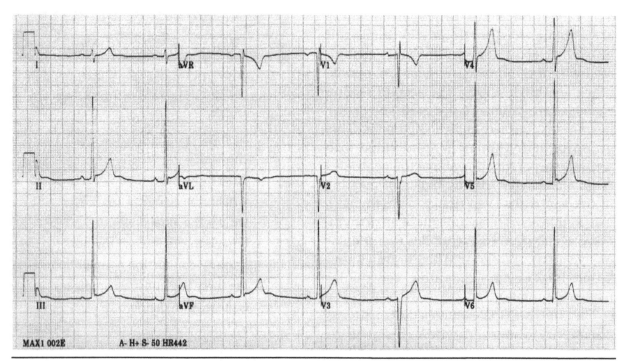

Figure 4.1 ECG demonstrating RAD. Notice the tall R waves in leads II and III with lead III being taller than lead II, and an RS complex with the S wave deeper than the R wave is tall is in lead I.

Identifying RAD and LAD

The following steps are a quick and easy way to identify axis deviation.

1. If the QRS complex in both lead I and aV$_F$ is positive, the axis is normal (i.e., –30° to +90°).

2. If the QRS in lead I is positive but in aV$_F$ is negative, the axis is leftward (i.e., somewhere between –90° and +60°), but left axis deviation exists only if the QRS in lead II is also negative (i.e., the axis is farther left than –30°).

3. If the QRS in aV$_F$ is positive and in lead I is negative, right axis deviation exists (i.e., the axis is between +90° and 180°).

4. If the QRS in both lead I and aV$_F$ is negative, extreme axis deviation exists (i.e., the axis lies between –90° and 180°), but commonly, this reading indicates lead transposition and not extreme axis deviation.

LAD

In general, if lead III shows a deep S wave, lead II shows an RS complex with the S wave deeper than the R wave is tall (or a QS complex), and leads I and aV$_L$ show a tall R wave, LAD is present (see figure 4.2).

Following are common causes of LAD:

- Normal variant (unusual)
- Left ventricular hypertrophy (not always seen in LVH)
- Left anterior hemiblock
- Left bundle branch block

Sinus Bradycardia

Normal heart rate is between 60 and 100 beats·min^{-1}. A reduction in heart rate below 60 beats·min^{-1} is defined as sinus bradycardia and is usually caused by a relative increase in vagal tone or a decrease in sympathetic tone (see figure 4.3 on p. 58). The rhythm is considered sinus bradycardia only when it arises wholly or predominantly from the sinus node (i.e., every QRS complex is preceded by a

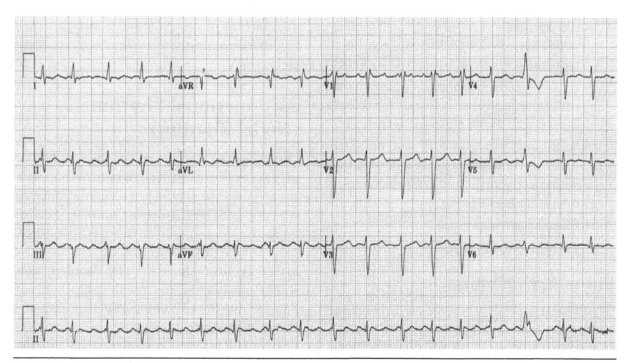

Figure 4.2 ECG demonstrating LAD. Lead III shows a deep S wave, lead II shows an RS complex with the S wave deeper than the R wave is tall (or a QS complex), and leads I and aV$_L$ show a tall R wave.

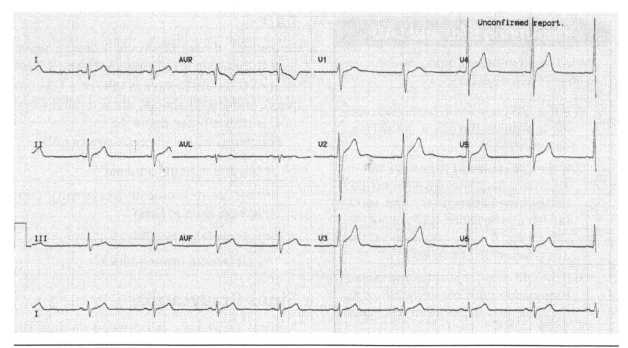

Figure 4.3   Sinus bradycardia.

P wave). Following are common causes of sinus bradycardia:

- Normal variant: Sinus bradycardia is common in normal people and is highly prevalent in people who are physically fit.

- Drug induced: A large number of drugs can increase vagal tone (e.g., digitalis) or decrease sympathetic tone (e.g., beta-blockers and calcium channel blockers).

- Hypothyroidism: Commonly associated with sinus bradycardia (hyperthyroidism is associated with sinus tachycardia).

- Sick sinus syndrome: A degenerative disease of the sinus node can result in sinus bradycardia and is more common in elderly people.

- Sleep apnea syndrome

- Carotid sinus syndrome

Sinus Arrhythmia

Beat-to-beat variability in sinus rate is common and results in small variations in the PP interval (the distance between consecutive P waves). The PP interval (or RR interval) should be constant with a difference of less than 0.16 s;

when it is greater than this, sinus arrhythmia is diagnosed (see figure 4.4). Sometimes termed sinus irregularity, this is an extremely common arrhythmia (particularly in children and young adults), and without evidence of heart disease, should be seen as a normal variant. The most common cause of sinus arrhythmia is changes in vagal tone during different phases of breathing (decreased vagal tone with inspiration and vice versa).

AV Junctional Rhythm (Nodal Rhythm)

Under certain circumstances the AV junction may act as the pacemaker. When this happens, the atria are stimulated in a retrograde fashion (the opposite direction from normal). Accordingly, the P wave will be inverted (i.e., the P wave will be positive in lead aV_R and negative in lead II). Although ventricular depolarization will be unaffected, the relationship between P waves and the QRS complex will be altered. If the atria depolarize prior to the ventricles, the P wave will be seen before the QRS complex and the PR interval will be short; if the two events occur simultaneously, the P wave may

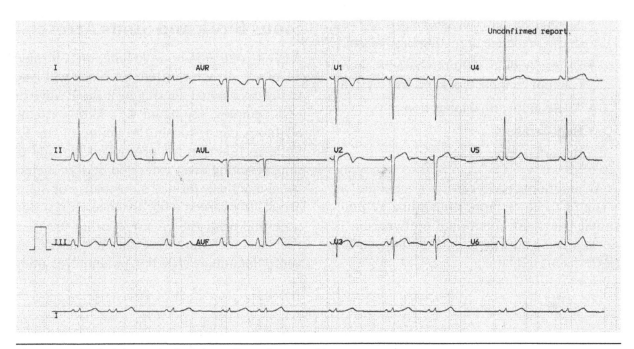

Figure 4.4   Marked sinus arrhythmia in an elite endurance athlete.

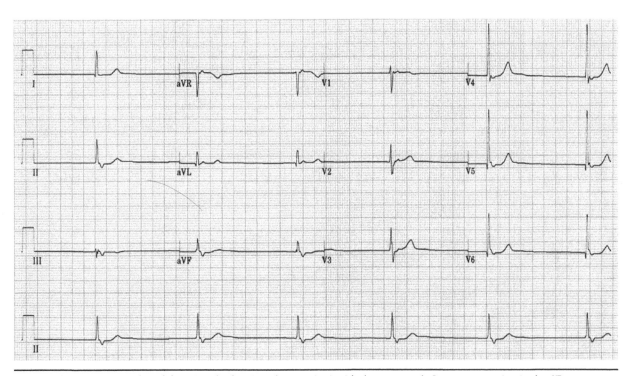

Figure 4.5   AV junctional/nodal escape rhythm (<60 beats·min⁻¹) with the retrograde P wave appearing in the ST segment.

be buried in the QRS complex; and if the atria depolarize after the ventricles, the P wave will be seen after the QRS complex (within the ST segment of the T wave). AV junctional rhythms are subdivided into AV junctional escape rhythms and AV junctional tachycardias.

AV junctional escape rhythms are defined as a consecutive run of AV junctional beats. The heart rate is characteristically slow (30 to 60 beats·min⁻¹; see figure 4.5). Following are common causes of AV junctional escape rhythms:

- Normal variant: Occasionally observed in the presence of resting bradycardia
- Drugs: Toxic reaction to beta-blockers, calcium channel blockers, and digitalis
- Acute myocardial infarction
- Hypoxemia
- Hyperkalemia

AV junctional tachycardias are defined as a run of three or more premature AV junctional beats. AV junctional tachycardia is analogous to paroxysmal atrial tachycardia (PAT, discussed later).

Sinus Block and Sinus Arrest

A failure of the SA node to function is termed SA block. An intermittent SA block will lead to the occasional missing beat (no P wave or QRS complex; see figure 4.6). More extreme SA block can result in the failure of the SA node to function for prolonged periods of time resulting in an extended period of electrical inactivity termed sinus pause or sinus arrest. To overcome the absence of an action potential from the SA node, other areas of the heart including the atria, AV junction, or ventricles initiate an action potential resulting

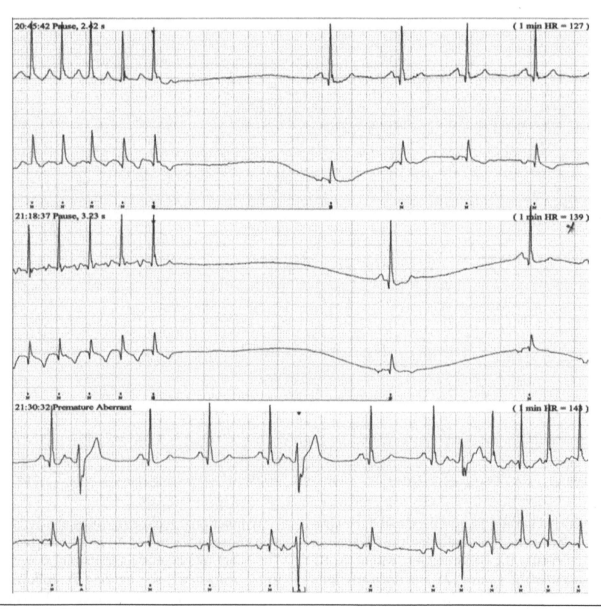

Figure 4.6 SA block.

in an escape beat. Following are causes of SA block and sinus arrest:

- Hypoxia
- Myocardial ischemia or infarction
- Hyperkalemia
- Digitalis toxicity
- Toxic reactions to other drugs including beta-blockers and calcium channel blockers

Degenerative disease of the SA node (sick sinus syndrome) can also lead to SA block and sinus arrest.

Atrioventricular Block (AV Block)

A disturbance in atrioventricular conduction leading to a transiently or permanently impaired transmission of the action potential through the AV junction is termed heart block. There are three categories, or degrees, of heart block ranging from the mildest (first-degree heart block) through second-degree heart block (subdivided into types I and II) to the most extreme (third-degree heart block, sometimes termed complete heart block). First- and second-degree AV block are termed incomplete heart block.

First-Degree AV Block

A constant prolongation of the PR interval >0.2 s is termed first-degree AV block (see figure 4.7). Following are common causes of first-degree AV block:

- Normal variant: First-degree AV block is commonly observed in highly trained athletes and likely reflects increased vagal tone. It is also occasionally observed in normal people.
- Drugs: Digitalis, beta-blockers, calcium channel blockers, and quinidine can depress AV function.
- Ischemic heart disease: First-degree AV block is common in inferior wall ischemia and infarctions because the right coronary artery supplies the AV junction as well as the inferior wall.
- Hyperkalemia
- Acute rheumatic fever

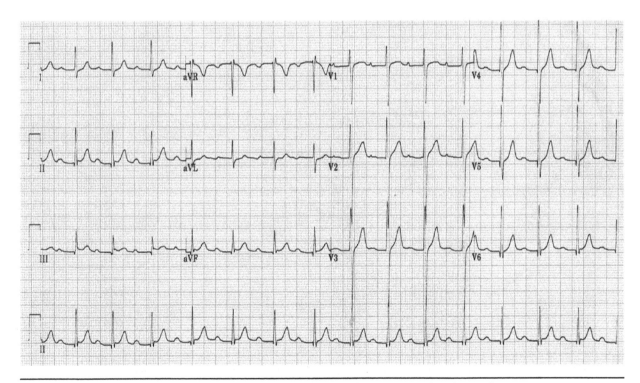

Figure 4.7   First-degree AV block. Notice the prolonged PR interval that is constant from beat to beat.

Note that the drugs mentioned, together with ischemic heart disease, are capable of causing all degrees of AV block.

Second-Degree AV Block

There are two forms of second-degree AV block: Mobitz type I (also called Wenckebach block) and Mobitz type II.

Mobitz type I (Wenckebach) AV block is characterized by a progressively increasing PR interval until the point at which the action potential is not conducted at all and the QRS complex is lost. This blocked beat is then followed by a relative recovery of the AV junction and a repeated pattern of progressive PR prolongation until a dropped beat occurs (see figure 4.8). The number of P waves occurring before the dropped beat can vary. Following are common causes of Mobitz type I (Wenckebach) AV block:

- Normal variant: Occasionally observed in highly trained athletes and likely reflects increased vagal tone.
- Drugs: Digitalis, beta-blockers, calcium channel blockers, and quinidine can depress AV function.

- Ischemic heart disease: Heart block is common in inferior wall ischemia and infarctions because the right coronary arteries supply the AV junction as well as the inferior wall.

Mobitz type II AV block is less common and more problematic than type I and indicates severe conduction system disease. Mobitz type II is characterized by an increased QRS duration (>0.12 s) due to the involvement of the His–Purkinje system, and a series of nonconducted P waves with intermittent appearances of QRS complexes (see figure 4.9).

A ratio is often assigned to second-degree AV block (e.g., 3:1 AV block indicates that every third P wave is followed by a QRS complex).

Third-Degree AV Block (Complete Heart Block)

In complete heart block there is no transmission of the action potential from the atria to the ventricles. Complete heart block is a serious and potentially life-threatening

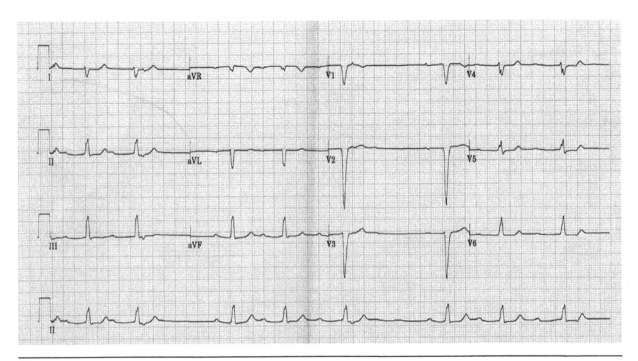

Figure 4.8   Mobitz type I (Wenckebach) second-degree AV block. Notice the progressively increasing PR interval until the QRS complex is dropped, followed by a relative recovery in the AV junction.

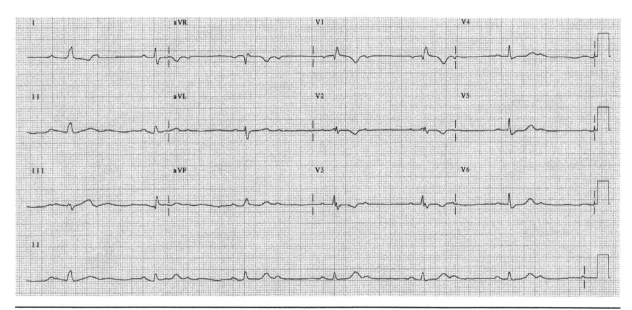

Figure 4.9 Mobitz type II second-degree AV block. Notice the wide QRS complexes together with a series of nonconducted P waves and the intermittent appearance of QRS complexes.

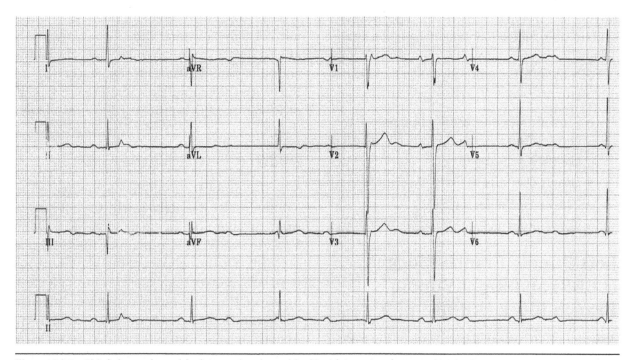

Figure 4.10 Third-degree heart block (complete heart block) with junctional pacemaker rhythm.

arrhythmia. The atria and ventricles are paced independently, and the ventricular action potential originates below the point of block in the AV junction. Third-degree heart block is characterized by a slow, fixed ventricular rate (30 to 60 beats·min⁻¹) and a regular atrial rate that is faster than the ventricular rate. There is no relationship between P waves and QRS complexes (complete heart block can occur in people in atrial flutter or fibrillation, who usually present with a slow, regular ventricular rate). QRS complexes can be of normal width (<0.12 s) if the action potential originates in the AV junction (junctional pacemaker; see figure 4.10) or abnormally wide (>0.12 s) resembling bundle branch block if the action

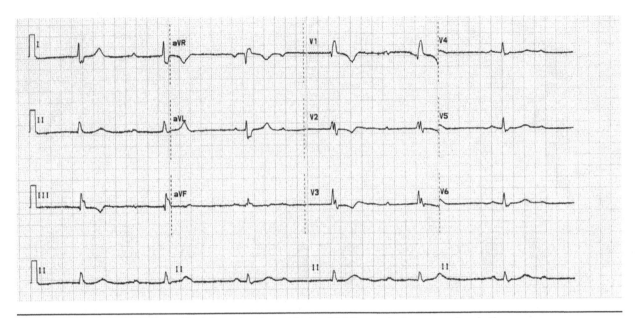

Figure 4.11   Third-degree heart block (complete heart block) with idioventricular pacemaker rhythm.

potential originates in the His–Purkinje system (idioventricular pacemaker; see figure 4.11).

Following are common causes of third-degree heart block:

- Chronic degenerative changes of the conduction system (commonly seen in geriatric populations)
- Drugs: Digitalis toxicity
- Lyme disease (arthritis)
- Acute myocardial infarction

Left Bundle Branch Block (LBBB) and Right Bundle Branch Block (RBBB)

A dysfunction resulting in a block in the transmission of the action potential along the right or left bundle branch is termed right bundle branch block (RBBB) or left bundle branch block (LBBB), respectively. A block in either bundle branch increases the duration of time for the depolarization of the ventricles leading to a number of characteristic changes on the ECG.

RBBB and LBBB can be permanent or intermittent. Furthermore, they may appear only at certain critical heart rates (i.e., are tachycardia or rate dependent; see chapter 5). An ECG with QRS complexes that are not typical of RBBB or LBBB indicates intraventricular delay.

RBBB

Septal depolarization will be unaffected in RBBB because it is activated by the left bundle branch. Therefore, a small septal R wave in V_1 and Q wave in V_6 remain in the presence of RBBB. The initial part of the simultaneous depolarization of the ventricles is unaffected; however, the later part is prolonged because right ventricular depolarization is delayed. Accordingly, a deep S wave (LV depolarization) is followed by an R' wave (RV depolarization) and a prolonged QRS complex in V_1. In V_6 the septal Q wave is followed by a tall R wave and an S' wave resulting in a prolongation of the QRS complex (>0.12 s; see figure 4.12). Occasionally, RBBB may be characterized by a tall notched R wave in V_1 (see figure 4.13). Secondary changes in the repolarization

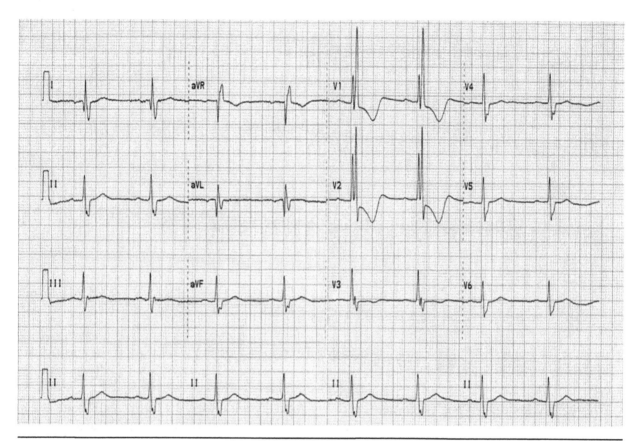

Figure 4.12 RBBB demonstrating prolongation of the QRS complex (>0.12 s) with rSR' in lead V_1 and qRS in V_6 and T-wave inversion in V_1 to V_3, II, III, and aV_F.

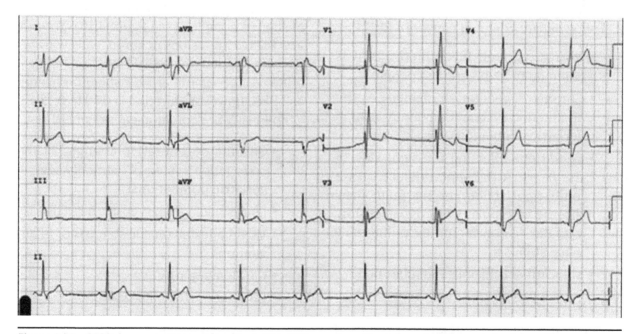

Figure 4.13 RBBB demonstrating prolongation of the QRS complex with a tall notched R wave in lead V_1 and T-wave inversion in V_1 to V_3, II, III, and aV_F.

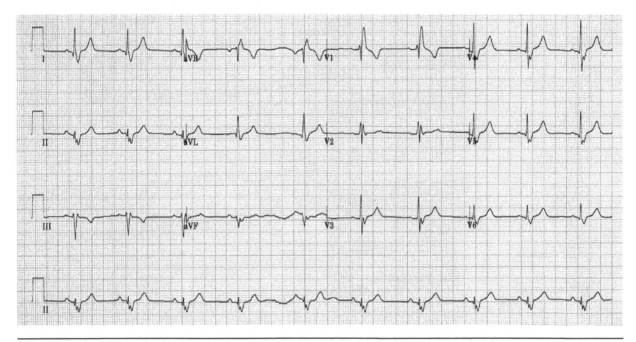

Figure 4.14   Incomplete RBBB demonstrating prolongation of the QRS complex (0.10-0.12 s) with rSR' in lead V_1 and qRS in V_6 and T-wave inversion in V_1 to V_3, II, III, and aV_F.

of the ventricles due to the prolongation of ventricular depolarization lead to T-wave inversion in the right precordial leads. Incomplete RBBB is characterized by an rSR' in V_1 and qRS in V_6 with a duration of 0.10 to 0.12 s (see figure 4.14).

Following are common causes of RBBB:

- Normal variant: Incomplete RBBB is common in highly trained people.

- Chronic degenerative changes of the conduction system: Commonly seen in geriatric populations.

- Conditions affecting the right side of the heart: Atrial septal defect with a left-to-right shunt; chronic pulmonary disease with pulmonary hypertension; valvular lesion including pulmonary valve stenosis.

- Myocardial ischemia and infarction

LBBB

In contrast to RBBB, LBBB affects septal and early ventricular depolarization. In LBBB the septum depolarizes from right to left resulting in a loss of the septal R wave in V_1 and septal Q wave in V_6. The QRS complex is character-istically wide (>0.12 s) with a QS complex in the right chest leads and an R wave in lead V_6. Occasionally, the QS wave in the right precordial leads shows notching leading to the characteristic W shape. The left precordial leads may also show notching resulting in an M-shaped complex (see figure 4.15). Secondary T-wave inversions occur in leads with tall R waves (e.g. left precordial leads). Incomplete LBBB has the same characteristic QRS changes as LBBB; however, duration is 0.10 to 0.12 s.

Following are common causes of LBBB:

- Chronic degenerative changes of the conduction system: Commonly seen in geriatric populations

- Long-standing hypertensive heart disease

- Valvular lesion (aortic stenosis or regurgitation)

- Cardiomyopathy

- Coronary artery disease

Fascicular Blocks

The left bundle branch (LBB) divides into two branches (fascicles or small bundles): the anterior fascicle and the posterior fascicle.

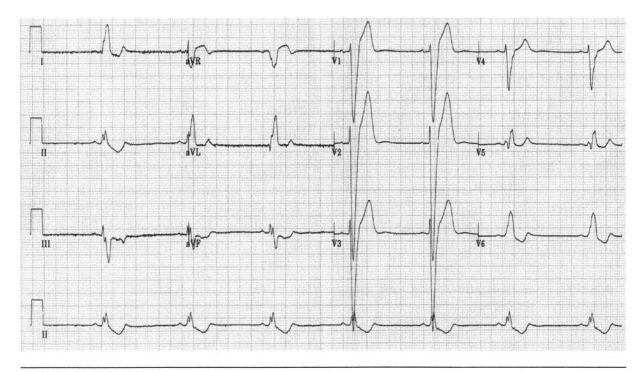

Figure 4.15 LBBB demonstrating prolongation of the QRS complex (>0.12 s) with a notched (W-shaped) QS wave in lead V$_1$ and a notched (M-shaped) R wave in V$_6$ and T-wave inversion in V$_1$ to V$_3$, II, III, and aV$_F$.

The right bundle branch (RBB) is composed of only one main fascicle. Accordingly, a block in the transmission of the action potential can occur at any point, or at multiple points, along the LBB before it bifurcates into the anterior and posterior fascicles, along each fascicle, or along the RBB. LBBB is a complete block of the main LBB, both fascicles, or both. If only one fascicle fails in the transmission of the action potential, the result is left anterior hemiblock or left posterior hemiblock.

Unlike in RBBB and LBBB, in hemiblock the QRS complex is not widened; instead, axis deviation is the characteristic marker (this is not normally observed in RBBB or LBBB).

• Left anterior hemiblock: Isolated left anterior hemiblock is diagnosed in the presence of a mean QRS axis >−45° (left axis deviation) and a QRS duration of <0.12 s. Diagnosis can only be made when other causes of left axis deviation have been excluded. Left posterior hemiblock occurs less frequently than LAFB (left anterior fascicular block) because of the anatomy of the posterior/inferior

fascicle and its superior blood supply compared to the anterior/superior fascicle.

• Left posterior hemiblock: Isolated left posterior hemiblock is diagnosed in the presence of a mean QRS axis >+120° (right axis deviation, RAD) and a QRS duration of <0.12 s. Other causes of RAD should be excluded (see discussion of right axis deviation) before the diagnosis of left posterior hemiblock can be made.

TACHYARRHYTHMIAS

Tachyarrhythmias are rhythms generally associated with elevated heart rates (>100 beats·min^{-1}) and fall under a number of categories including a heart rate >100 beats·min^{-1} originating from the sinus node (sinus tachycardia); an elevated heart rate as a result of an arrhythmia originating in the atria or AV junction (supraventricular arrhythmias including PAC, PAT, AV junctional rhythms, SVT, and AF); arrhythmias originating in the ventricles (ventricular arrhythmias including isolated ventricular ectopics [PVCs], bigeminy, trigeminy, couplets, and triplets); an elevated heart rate as a result of an arrhythmia originating in the ventricles (ventricular tachycardia including torsades de pointes, R-on-T, and ventricular fibrillation); and an elevated heart rate without an obvious origin (idiopathic ventricular tachyarrhythmia including RVOT-VT and CPVT). Supraventricular versus ventricular arrhythmias are presented in table 4.1. The following sections examine each of these conditions with supporting ECGs to illustrate their presentations.

Supraventricular Arrhythmias

This section addresses the supraventricular arrhythmias sinus tachycardia and PAC.

Sinus Tachycardia

Normal heart rate is between 60 and 100 beats·min^{-1}. A heart rate greater than 100 beats·min^{-1} is defined as sinus tachycardia and is usually caused by a relative increase in sympathetic tone or a decrease in vagal tone (see figure 4.16). The rhythm is sinus tachycardia only when it arises

Table 4.1   Characterization of Supraventricular Versus Ventricular Arrhythmia

	Supraventricular (normal conduction)	Supraventricular (abnormal conduction)	Ventricular
QRS complex duration	<0.12 s	≥0.12 s	≥0.12 s
Nomenclature	Normal	Abnormal: same initial vector as normal beats and pattern often triphasic	Abnormal: pattern may be biphasic or triphasic
P wave	Present or absent	Present or absent	Absent
Rhythm	No compensatory pause	No compensatory pause	Often followed by a compensatory pause

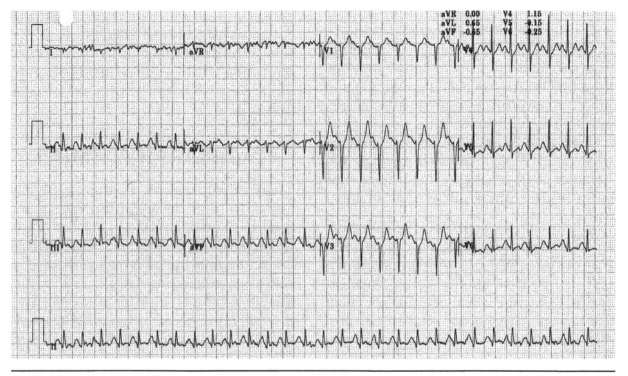

Figure 4.16   Sinus tachycardia.

wholly or predominantly from the sinus node (i.e., every QRS complex is preceded by a P wave). In sinus tachycardia each QRS complex is preceded by a P wave with a regular PR interval (the P wave may be superimposed on the preceding T wave making identification difficult). Following are common causes of sinus tachycardia:

- Normal variant: Anxiety, emotion, and exertion or exercise
- Drug induced: A large number of drugs can increase sympathetic tone (e.g., Adrenalin, tricyclic antidepressants, isoproterenol, cocaine). Some drugs can decrease or block vagal tone (e.g., atropine).
- Hyperthyroidism
- Congestive heart failure (an increased sympathetic tone is commonly observed with pulmonary edema)
- Pulmonary embolism (sinus tachycardia is the most common arrhythmia in someone with a pulmonary embolism)
- Acute myocardial infarction
- Hypotension and shock
- Alcohol withdrawal

Premature Atrial Contraction (PAC)

PACs, sometimes termed atrial ectopics (AE), are ectopic (nonsinus) beats arising somewhere within the left or right atrium (see figure 4.17). The following will help in identifying PAC.

Characteristics of a PAC

- The beat is premature and occurs before the next beat is expected (in contrast, escape beats occur after a pause in the expected rhythm).
- The PAC is usually indicated by a P wave that has a slightly different PR interval and shape compared with the normal sinus beat (occasionally the P wave is buried in the preceding T wave and therefore not visible).
- Following a PAC, there is usually a pause prior to the commencement of sinus rhythm.
- Usually, the QRS complex is entirely normal (occasionally, a PAC may cause aberrant ventricular conduction resulting in a widened QRS).

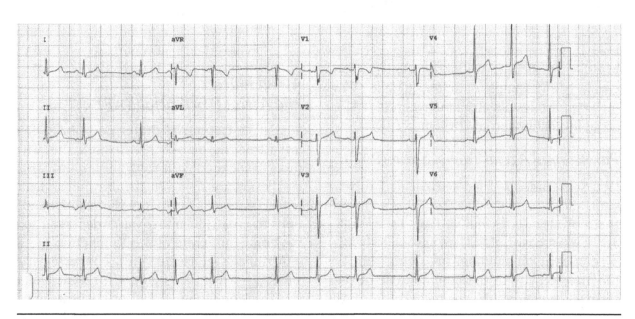

Figure 4.17 Premature atrial contractions (PACs). Notice the early appearance of the P wave with a slightly different shape and PR interval, and the compensatory pause until the next sinus beat.

If the PAC occurs very early during the AV junction refractory period (see chapter 1), the PAC will be blocked and no QRS will be present. A pause will follow the blocked PAC leading to an irregular rhythm. PACs may occur frequently or sporadically. Two consecutive PACs are termed paired PACs, and the occurrence of a PAC after each sinus beat is termed atrial bigeminy.

PACs are very common in both normal and diseased hearts; however, they may be the forerunner of paroxysmal atrial tachycardia (PAT) or atrial fibrillation (AF). Following are common causes:

- Normal variant: Associated with emotional stress and excessive caffeine ingestion.

- Heart disease: Any form of heart disease can cause PACs

- Drugs: Sympathomimetic drugs (e.g., adrenaline, isoproterenol, theophylline)

AV Junctional Rhythms

Supraventricular arrhythmias like SVT, PAT, and AF often originate in the AV junction. This section addresses the arrhythmias that originate in the AV junction.

Supraventricular Tachycardia (SVT)

An SVT is defined as a tachycardia in which the pacemaker is located above the ventricles in either the atria or AV junction. Sinus tachycardia is an example of an SVT. Other SVTs include paroxysmal atrial tachycardia (PAT), atrial flutter, and atrial fibrillation (AF).

Paroxysmal Atrial Tachycardia (PAT)

PAT, sometimes termed paroxysmal supraventricular tachycardia (PSVT), is a run of three or more consecutive PACs (see figure 4.18). Most cases of PAT are associated with a reentrant mechanism involving the AV junction leading to the common use of the term *AV junctional/nodal tachycardia*. PAT

Characteristics of PAT

- Heart rate is usually regular and between 140 and 250 beats·min^{-1}.

- P waves may or may not be visible and are usually a different shape with a short-, normal-, or long-duration PR interval.

- QRS is normal (a wide QRS may be observed in patients with LBBB or rate-dependent LBBB).

can occur for brief periods of time (seconds) or be sustained over long periods (weeks). Characteristics of PAT are listed on this page.

The higher heart rate and regularity (constant RR interval) observed in PAT distinguishes it from other forms of supraventricular tachycardia. Differentiating PAT with wide QRS complexes (e.g., LBBB) can be very difficult.

Following are common causes of PAT:

- Normal variant: Associated with emotional stress or excessive caffeine or alcohol ingestion.

- Heart disease: Any form of heart disease can cause PACs.

- Ventricular preexcitation: Wolff-Parkinson-White (WPW) syndrome or Lown-Ganong-Levine (LGL) syndrome.

Atrial Flutter and Atrial Fibrillation (AF)

Similar to PAT, atrial flutter and fibrillation arise from an ectopic (nonsinus) site. The rates associated with atrial flutter and atrial fibrillation are 250 to 350 beats·min^{-1} and 400 to 600 beats·min^{-1}, respectively. In both conditions the ventricular rate varies depending on the conduction rate of the AV junction.

Atrial flutter results in a characteristic sawtooth wave (see figure 4.19). Ventricular rates are commonly 75, 100, or 150 beats·min^{-1}

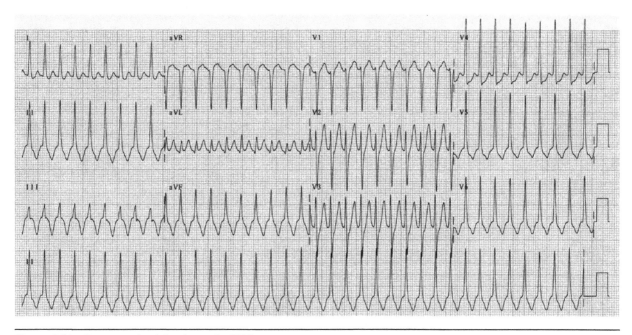

Figure 4.18 Paroxysmal atrial tachycardia (PAT).

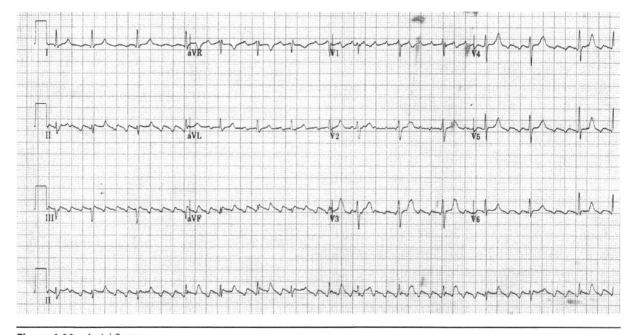

Figure 4.19 Atrial flutter.

associated with a 4:1, 3:1, or 2:1 conduction of the atrial rate, respectively (a 1:1 flutter is very uncommon because of the limitations of AV junction conduction). Atrial flutter is rarely seen in the normal heart and is common in all forms of heart disease.

Atrial fibrillation (AF) is one of the most commonly seen arrhythmias. It produces a characteristically irregular pattern of fibrillatory waves (F waves) that can be fine or coarse (see figure 4.20). The random stimulation of the AV junction leads to an

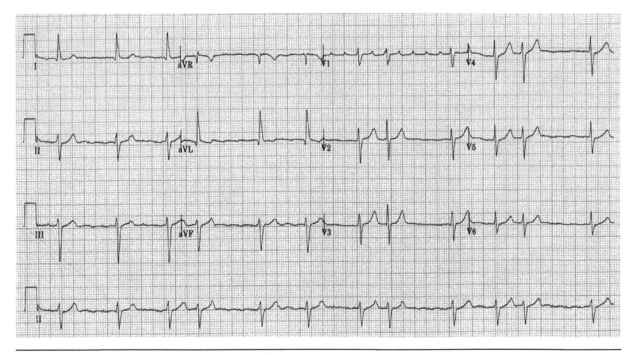

Figure 4.20   Atrial fibrillation demonstrating both fine and coarse F waves.

irregular depolarization of the ventricles and resultant RR interval. Ventricular rates in patients with a normal AV junction are 110 to 180 beats·min⁻¹. AF may be paroxysmal, lasting minutes, weeks, or days; or chronic, persisting for months, years, or permanently. AF can occur in normal hearts as a result of emotional stress or excessive alcohol ingestion (holiday heart syndrome). AF is commonly seen in all forms of organic heart disease particularly those leading to atrial dilatation and fibrosis, including chronic ischemic heart disease, hypertension, and valvular (mitral) heart disease. Other causes include hyperthyroidism, chronic pericarditis, pulmonary emboli, cardiomyopathies, and cardiac surgery.

Ventricular Arrhythmias

The arrhythmias in this section originate in the ventricles.

Ventricular Ectopics (VEs)

VEs, sometimes termed premature ventricular contractions (PVCs), occur before the next normal beat is expected and result in an abnormally wide (>0.12 s) QRS complex that has an abnormal, aberrant appearance compared with normal QRS complexes in the same lead (see figure 4.21). The frequency of VEs can vary from a single isolated aberrant beat to many occurring in various combinations. Two consecutive VEs are termed couplets (see figure 4.22), three consecutive VEs are triplets (see figure 4.23), and a run of three or more VEs is termed ventricular tachycardia (VT). A VE that occurs in a rhythmical fashion following every normal beat is termed bigeminy (see figure 4.24), and one following two normal beats is termed trigeminy (see figure 4.25).

VEs are normally followed by a compensatory pause. In some cases the VE may occur

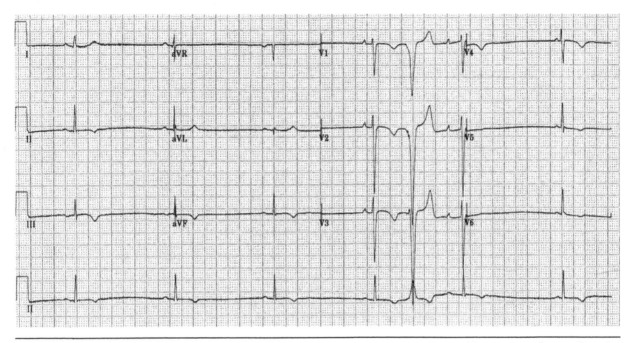

Figure 4.21 An isolated ventricular ectopic (VE) occurring prior to the next expected normal beat with an abnormally wide (>0.12 s), aberrant QRS complex.

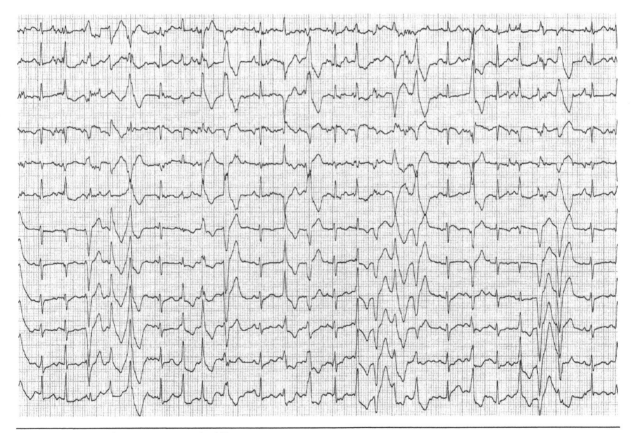

Figure 4.22 Couplets: two consecutive VEs.

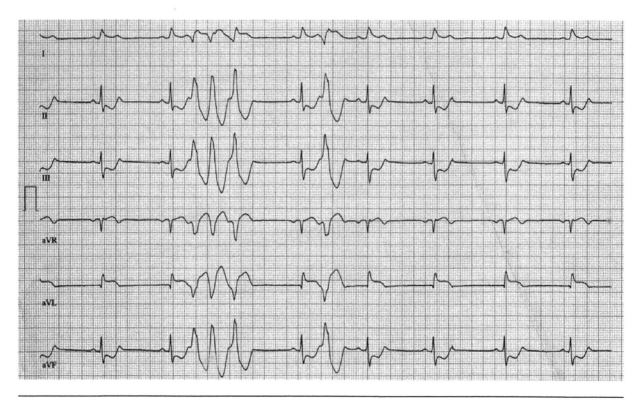

Figure 4.23 Triplets: three consecutive VEs.

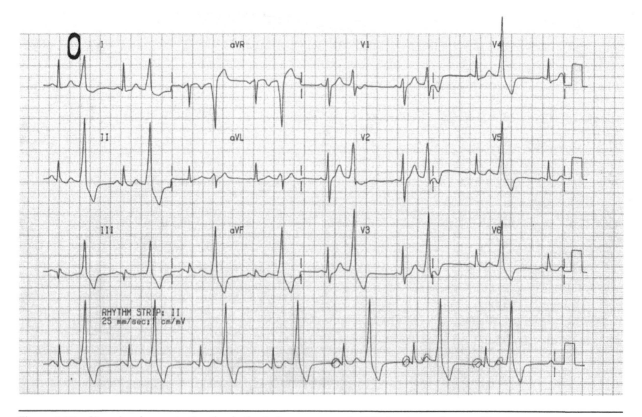

Figure 4.24 Bigeminy demonstrating the rhythmical appearance of a VE following each normal beat.

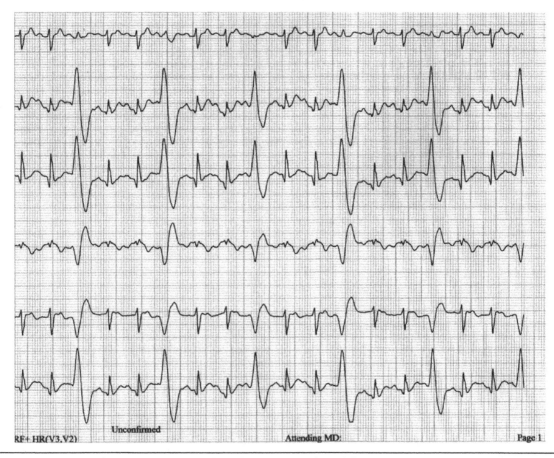

RF+ HR(V3,V2)   Unconfirmed   Attending MD: Page 1

Figure 4.25 Trigeminy demonstrating the rhythmical appearance of a VE following every two normal beats.

near the peak of the preceding T wave, which is termed the R-on-T phenomenon. The appearance of R-on-T is important to note because it may precipitate ventricular tachycardia or ventricular fibrillation. The aberrant shape of VEs may be uniform (monomorphic) or multiform in a single lead (VEs have different shapes in different leads) depending on the location of the ectopic focus (see figure 4.26).

Following are common causes of VEs:

- Normal variant: VEs are very common in normal hearts. Anxiety and excessive caffeine ingestion may cause VEs.

- Drugs: Adrenalin, isoproterenol, aminophylline; digitalis toxicity

- Heart disease: Any form of heart disease can cause VEs.

- Electrolyte disturbances: Hypokalemia, hypomagnesemia

- Lung disease

- Hypoxia

Ventricular Tachycardia (VT)

Ventricular tachycardia is defined as a run of three or more consecutive VEs (see figure 4.27). VT may occur as a single, isolated burst, may recur paroxsymally, or may be sustained (>30 s). Heart rate is usually 100 to 200 beats·min⁻¹. Sustained VT is a life-threatening arrhythmia associated with an induced hypotension and possible degradation to ventricular fibrillation (VF). Common causes of VT are the same as those for VEs; however, VT rarely occurs in the absence of heart disease. VT may occur

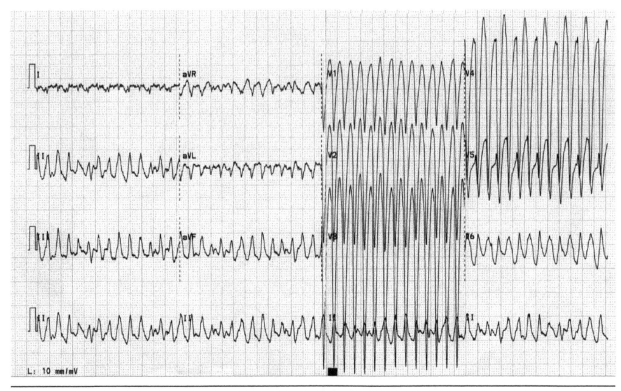

Figure 4.26   Multiform VEs in a single lead demonstrating a multifocal origin of the VEs.

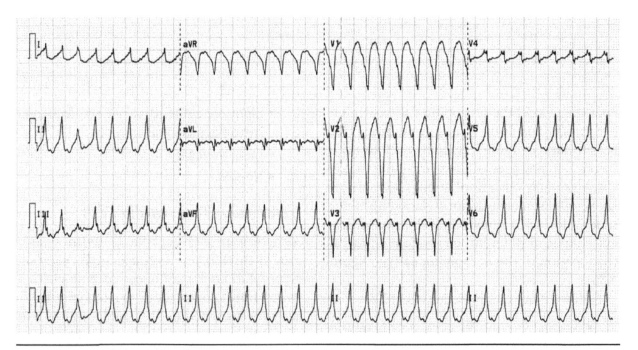

Figure 4.27   Ventricular tachycardia (VT).

in structurally normal hearts and is called idiopathic ventricular tachyarrhythmia; it includes right ventricular outflow tract VT (RVOT-VT; see case study 4 in chapter 7) and catecholaminergic polymorphic VT (CPVT).

A particular form of VT termed torsades de pointes is characterized by a continually changing shape of the QRS complex indicating a variable/circular focus of ectopic origin (see figure 4.28). Following are common causes of torsades de pointes:

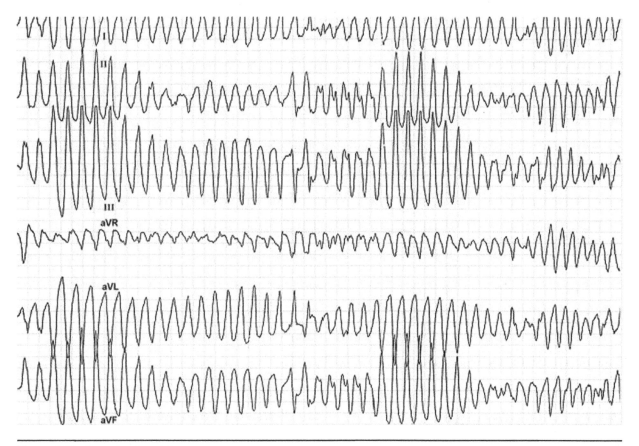

Figure 4.28 Torsades de pointes.

- Drug toxicity: Antiarrhythmic agents including quinidine, disopyramide, procainamide, and psychotropic drugs including tricyclic antidepressants and phenothiazines

- Electrolyte disturbance: Hypokalemia, hypomagnesemia, hypocalcemia

- Long QT syndrome

- Myocardial ischemia

Ventricular Fibrillation (VF)

VF results in the ventricles quivering in an asynchronous fashion leading to a complete loss of cardiac output, blood pressure, and consciousness. The ECG shows characteristic fine or coarse fibrillatory waves (see figure 4.29 on p. 78). VF is a life-threatening arrhythmia requiring immediate cardioversion (defibrillation). VF can occur in patients with heart disease of any type and may occur

in the normal heart as a result of drug toxicity (e.g. Adrenalin, cocaine).

ATRIAL AND VENTRICULAR ENLARGEMENT

Left atrial enlargement (LAE), right atrial enlargement (RAE), and biatrial enlargement can be observed on the 12-lead resting ECG and are characterized by changes in P-wave duration and morphology. Left ventricular enlargement (LVH), right ventricular enlargement (RVH), and biventricular enlargement can be observed on the resting 12-lead ECG and are characterized by changes in the QRS complex, ST segment, and T-wave duration and morphology. Often, ventricular enlargement is accompanied by atrial enlargement with characteristic features on the ECG. This section examines each of these conditions and provides supporting ECGs to illustrate their presentations.

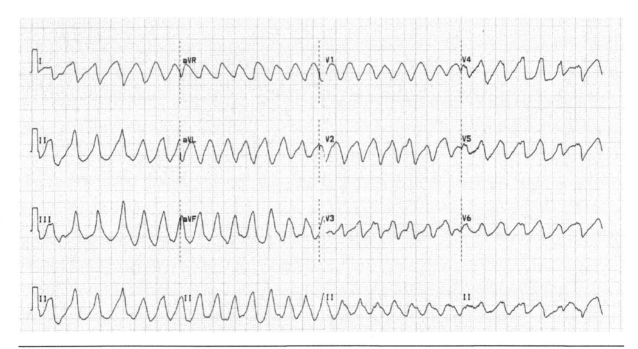

Figure 4.29   Ventricular fibrillation (VF).

Atrial Enlargement

Dilatation of the chamber or hypertrophy of the cardiac muscle surrounding the chamber results in characteristic changes on the ECG. Dilatation and hypertrophy often occur simultaneously and are commonly caused by pressure or volume overload. Exceptions to this are the heart muscle diseases including hypertrophic cardiomyopathy (HCM) and dilated cardiomyopathy (DCM).

Right Atrial Enlargement (RAE)

RAE causes an increase in the electrical activity resulting in a greater P-wave voltage (amplitude) and a shift of the P-wave axis to a more rightward position, often toward a vertical axis. In RAE the P wave is abnormally tall (>0.25 mV) with a normal duration (<0.12 s), and is often termed P-pulmonale (RAE is often observed in pulmonary disease). Tall, narrow P waves are often observed in leads II, III, aV$_F$, and V$_1$ (see figure 4.30).

Following are common causes of RAE:

- Acute pulmonary disease: Bronchial asthma and pulmonary embolism
- Chronic pulmonary disease: Emphysema and bronchitis
- Congenital heart disease: Pulmonary valve stenosis, atrial septal defects, and tetralogy of Fallot

Left Atrial Enlargement

LAE increases the time required for the action potential to spread across the atria resulting in an abnormally wide (>0.12 s) P wave that has either a normal or increased amplitude. The delayed depolarization can result in a humped or notched P wave commonly termed P mitral (this term technically refers only to a humped or notched P wave that is the result of mitral valve stenosis). Occasionally, a biphasic P wave is observed in lead V$_1$ with an initial positive deflection and a wide (>0.04 s), deep (>0.1 mV) terminal portion (see figure 4.31).

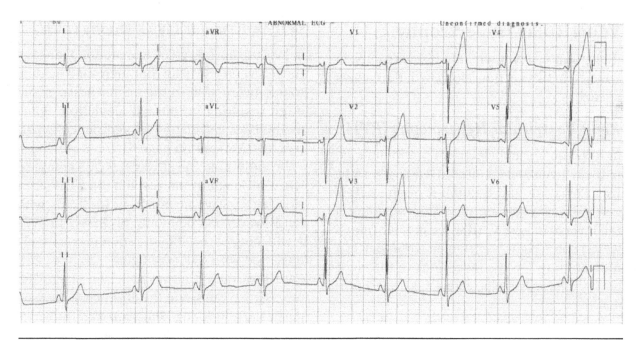

Figure 4.30 Right atrial enlargement (RAE). Notice the tall (>0.25 mV), narrow (<0.12 s) P waves in leads II, III, aV$_F$, and V$_1$.

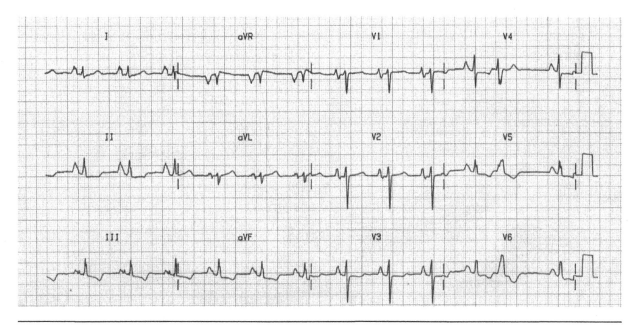

Figure 4.31 Left atrial enlargement (LAE) is characterized by wide (>0.12 s) bifid P waves in lead II, a biphasic P wave in lead V$_1$, or both. Right atrial enlargement (RAE) is characterized by a peak P wave (>0.25 mV) in leads II, V$_1$, or both.

Following are common causes of LAE:

- Congestive heart failure associated with all causes including ischemic heart disease and cardiomyopathy

- Valvular heart disease: Aortic stenosis, aortic regurgitation, and mitral stenosis
- Hypertensive heart disease

Ventricular Hypertrophy

Right and left ventricular hypertrophy is covered in this section, along with the causes of each.

Right Ventricular Hypertrophy (RVH)

In RVH the normal electrical predominance of the left ventricle is altered leading to right axis deviation and a number of characteristic QRS changes on the ECG, including a tall R wave in lead V_1 (R wave greater than S wave) and a right ventricular strain pattern indicated by T-wave inversions in the right and middle precordial leads (V_1-V_3; see figure 4.32).

Following are common causes of RVH:

- Congenital heart disease: Pulmonic stenosis, atrial septal defect, tetralogy of Fallot

- Chronic pulmonary disease

- Mitral stenosis

Left Ventricular Hypertrophy (LVH)

In LVH the normal electrical predominance of the left ventricle is further increased leading to a leftward shift in the mean QRS axis and often resulting in left axis deviation. Some methods for calculating left ventricular hypertrophy (LVH) are on page 81.

In addition to the size of the QRS complex, left ventricular strain resulting in ST depression and T-wave inversion in the leads with tall R waves is commonly observed in LVH. A widening of the QRS duration is commonly observed with LAE (see figure 4.33 on p. 82). Care is warranted, however, because large QRS waves are often see in the absence of disease, particularly in young, slim people and athletes.

Following are common causes of LVH:

- Valvular heart disease: Aortic stenosis, aortic regurgitation, and mitral regurgitation

- Systemic hypertension

- Cardiomyopathy

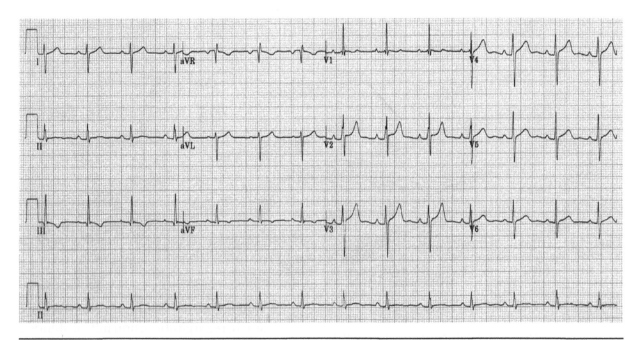

Figure 4.32   Right ventricular hypertrophy (RVH). Notice the tall R wave in V_1 and a right ventricular strain pattern indicated by T-wave inversions in leads V_1 through V_3.

Calculating Left Ventricular Hypertrophy (LVH)

- LVH is present when an R wave >35 mm (3.5 mV) occurs in any lead. Electrical criteria for LVH may be considered when the R-wave in any lead (actually V_4-V_6) is >25 mm (>2.5 mV) (and alternatively if the S-wave in V_1-V_3 is >25 mm) (Framingham criteria).

- The Sokolow criteria use the sum of the amplitude of the S wave in V_1 and R wave in V_5 or V_6 (the lead with the largest wave is used in the calculation). LVH is present when the S wave in V_1 + the R wave in V_5 or V_6 > 35 mm (3.5 mV). Alternatively, an R wave in aV_L > 11 mm (1.1 mV).

- The Romhilt-Estes point score system: Probable LVH is diagnosed with a point score of 4 points, and definite LVH is diagnosed with a point score of 5 or more (table 4.2).

- Cornell voltage criteria: These more recent criteria are based on echocardiographic correlative studies designed to detect a left ventricular mass index >132 g/m² in men and >109 g/m² in women.

- For men: S in V_3 + R in aV_L > 2.8 mV (28 mm)

- For women: S in V_3 + R in aV_L > 2.0 mV (20 mm)

Table 4.2 Romhilt-Estes Criteria for the Determination of LVH From the Resting 12-Lead ECG*

Criterion	Points
Any limb R wave or S wave ≥2.0 mV (20 mm) or S in V_1 or V_2 ≥3.0 mV (30 mm) or R in V_5 or V_6 ≥3.0 mV (30 mm)	3
ST- and T-wave changes typical of LVH Taking digitalis Not taking digitalis	1 3
Left atrial abnormality P terminal force in V_1 is ≥1 mm in depth with a duration of ≥0.04 s	3
Left axis deviation ≥–30°	2
QRS duration ≥90 ms	1
Intrinsicoid deflection in V_5 or V_6 ≥0.05 s (interval between the beginning of the QRS interval and the peak of the R wave)	1

*Probable LVH is diagnosed with a score of 4 points, and definite LVH is diagnosed with a score of 5 points or more.

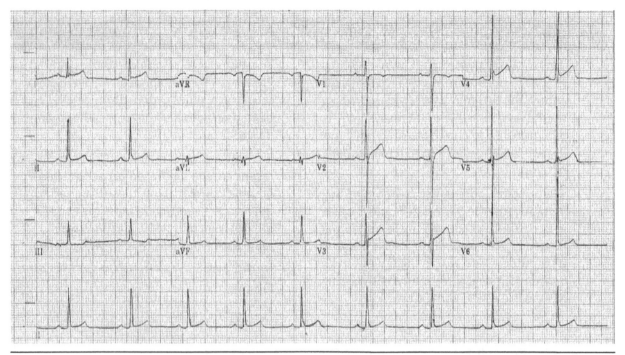

Figure 4.33   Left ventricular hypertrophy (LVH).

ECG ABNORMALITIES IN DISEASES OF THE CORONARY CIRCULATION: ISCHEMIA AND MYOCARDIAL INFARCTION

Cardiovascular disease (CVD) leading to impaired blood flow through the coronary vasculature can result in myocardial ischemia that may evolve leading to myocardial infarction. Myocardial ischemia is a reversible condition resulting from an oxygen demand that is greater than oxygen supply. In contrast, myocardial infarction (MI) results in the necrosis (death) of myocardial cells. Both ischemia and infarction can occur in distinct regions of the ventricular myocardium localized to the subendocardial border or encompassing the entire width of the ventricular myocardium (transmural; see figure 4.34). The following section examines transmural myocardial infarction and associated ECG changes and the identification of infarct location from the 12-lead ECG. This is fol-

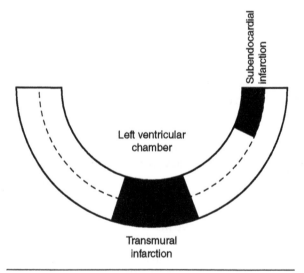

Figure 4.34   Cross section of the ventricular wall demonstrating transmural (full thickness) and subendocardial (inner half) regions.

lowed by an examination of subendocardial ischemia and infarction.

Transmural Infarction

Myocardial ischemia leading to necrosis of a section of the entire thickness of the ventricular wall results in depolarization (QRS)

Characteristic ST-T Changes in Transmural MI

- Acute phase: ST-segment elevation and tall, positive (hyperacute) T waves. The shape of the ST-segment elevation is variable (see figure 4.35). ST-segment elevation is characteristic of acute transmural infarction and as a consequence is termed ST-elevation myocardial infarction (STEMI).

- Evolving phase: Deep T-wave inversions in those leads that previously demonstrated ST-segment elevation.

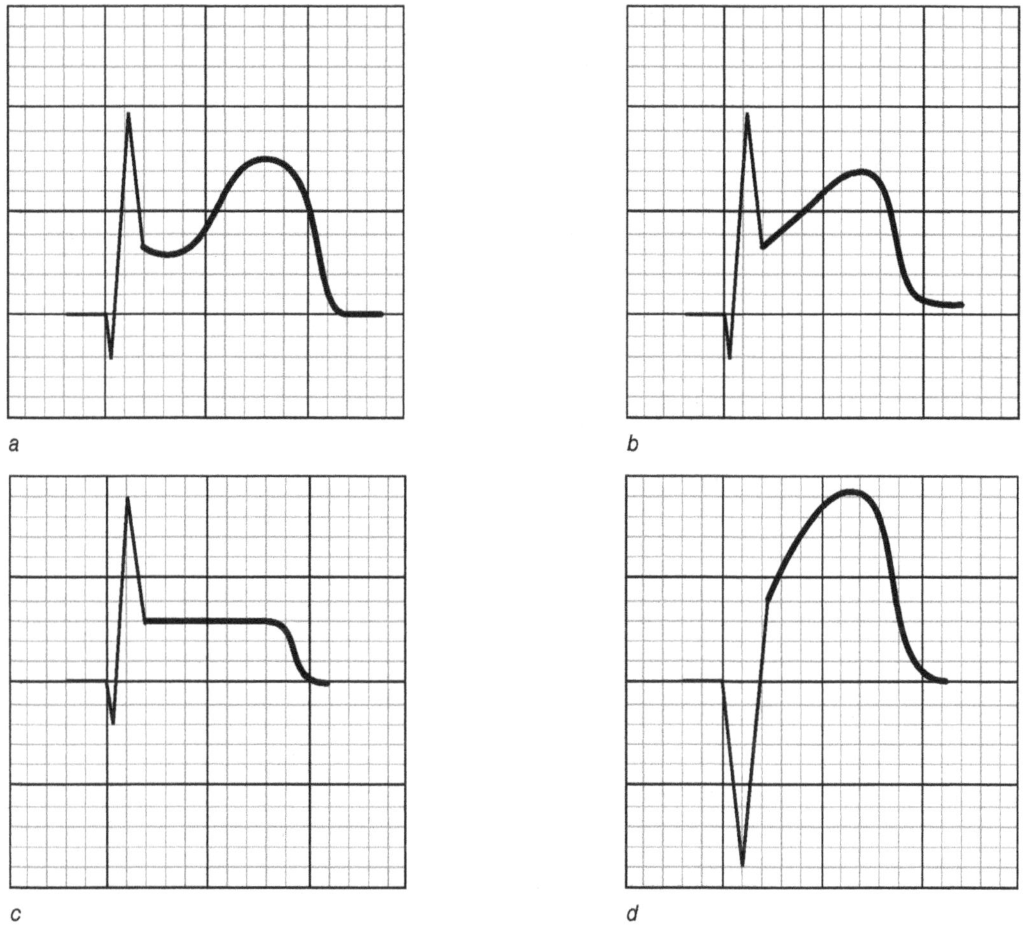

Figure 4.35   The variable shape of ST-segment elevation observed in acute MI.

and repolarization (ST-T) changes in various stages reflecting the progression of the condition.

ST-T Changes in Transmural MI

The two key stages of ST-T changes are acute (minutes to hours) and evolving (hours to days). Both result in the following characteristic changes.

ECG changes in certain leads indicate the location of the transmural MI. For example,

acute anterior wall MI results in ST-segment elevation and hyperacute T waves in the anterior leads (V_1-V_6, I, and aV_L; see figure 4.36). Of note, the anterior and inferior leads commonly demonstrate inverse, or reciprocal, patterns; for example, the ST-segment elevation and hyperacute T waves observed in the anterior leads (V_1-V_6, I, and aV_L) of an acute anterior MI are combined with ST-segment depression in the inferior leads (II, III, and aV_F; see figure 4.37).

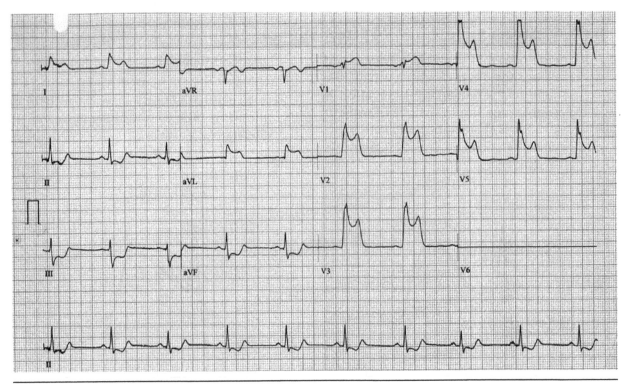

Figure 4.36 Acute anterior MI demonstrating ST-segment elevation and hyperacute T waves in the anterior leads (V_1-V_6, I, and aV_L) combined with reciprocal ST-segment depression in the inferior leads (II, III, and aV_F).

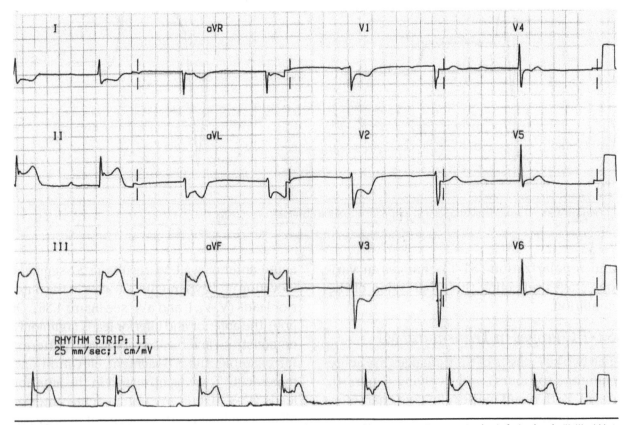

Figure 4.37 Acute inferior MI demonstrating ST-segment elevation and hyperacute T waves in the inferior leads (II, III, aV_F) combined with reciprocal ST-segment depression in the anterior leads (V_1-V_6, I, and aV_L).

Table 4.3 Identifying and Locating an Acute Transmural Infarction From the Presence of ST-Segment, T Wave, Q Wave, and Other ECG Changes on the Resting 12-Lead ECG

	Infarct location	ST elevation and hyperacute T waves	ST depression	Significant Q waves	Other ECG findings
Left ventricular MI	Anteroseptal			V_1-V_2	Loss of R wave in V_1 and V_2, with QS complex
	Anterior	V_2-V_4		V_2-V_4	Loss of normal R-wave progression
	Anterolateral	V_5-V_6		V_5-V_6	
	Extensive anterior	V_1-V_6, I, aV_L	II, III, aV_F	V_1-V_6	Loss of normal R-wave progression
	High lateral			V_5-V_6, I, aV_L	
	Inferior	II, III, aV_F	V_1-V_6, I, aV_L	II, III, aV_F	
	Posterior	ST depression V_1-V_2			R > S in V_1 (posterior may merge with inferior or lateral wall infarction)
Right ventricular MI	Right ventricle	V_1-V_2			

ST-segment elevation and hyperacute T waves are only usually present during the acute phase (minutes to hours). Persistent ST-segment elevation for days to weeks may be indicative of ventricular aneurysm.

QRS Changes in Transmural MI

A characteristic ECG sign of a transmural MI is the presence of new Q waves indicating the loss of electrical voltages due to necrosis. As with ST-T changes, the presence of Q waves in certain leads can be used to locate the MI (see table 4.3). Identifying Q waves can be problematic. Small, narrow (<0.04 s) Q waves are commonly observed in the left precordial leads (V_4-V_6) and in one or more of leads I, II, III, aV_L, and aV_F. Abnormal Q waves are usually >0.04 s in duration. Q waves are not always indicative of acute MI. Noninfarctional Q waves occur in a variety of pathologies, including LVH, LBBB, chronic lung disease,

and DCM. Over time (months to years) Q waves may persist but become smaller and, rarely, may disappear.

Subendocardial Ischemia

The subendocardium is most distant from the coronary blood supply and closest to the high ventricular pressures. For this reason, the subendocardium is particularly susceptible to ischemia. Subendocardial ischemia is characterized by ST-segment depression (ST-segment elevation may be seen in aV_R) that is transient (reversible) in nature and may be associated with stressors that cause an increased myocardial oxygen demand (e.g., exercise, stress, and exposure to the cold). Subendocardial ischemia is often accompanied by symptoms of chest, neck, or arm pain and shortness of breath termed angina pectoris. (For a description of subendocardial ischemia and exercise testing, see chapter 5.) Unlike ST-segment elevation, ST-segment depression does not indicate the location of the affected myocardial area the way ST-segment elevation indicates transmural MI.

Some patients suffer from a condition termed Prinzmetal's angina. Unlike those with subendocardial ischemia, these patients demonstrate transient ST-segment elevation during angina attacks, which occur at rest and at night (unlike the stress-induced angina pectoris observed with subendocardial ischemia). Prinzmetal's angina is associated with coronary artery spasm with or without coronary artery disease (see figure 4.38).

Subendocardial Infarction

In the presence of subendocardial infarction, Q waves are not commonly observed and ST-segment depression is persistent (see figure 4.39). Accordingly, the term *non-ST elevation MI (NSTEMI)* is used to describe subendocardial infarction. Formerly, subendocardial infarction was termed *non-Q wave infarction,* but because rapid revascularization of an ST-elevation MI (STEMI) can lead to aborted Q waves, altering diagnosis, the condition was renamed. Indeed, the presence or absence of ST-segment elevation (i.e., STEMI versus NSTEMI), rather than the presence or absence of Q waves (which arise later) guides the early treatment of acute coronary syndromes. The presence of T-wave inversion without ST-segment depression is occasionally observed in NSTEMI (see figure 4.40). Figure 4.41 on page 88 summarizes the common ECG changes observed in the presence of various forms of myocardial ischemia and infarction.

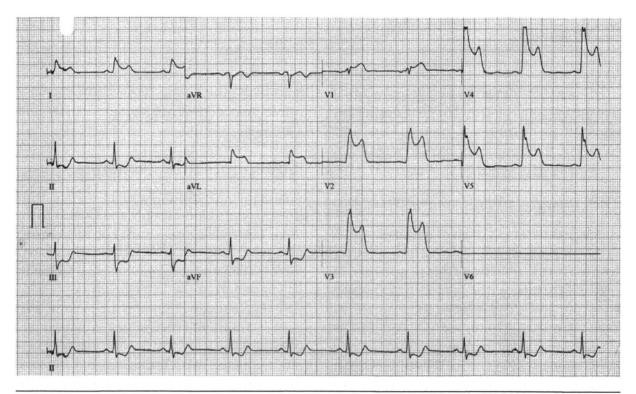

Figure 4.38   An ECG from a patient suffering from Prinzmetal's angina. Notice the ST-segment elevation similar to that observed in transmural MI. These changes, however, are transient and reversible.

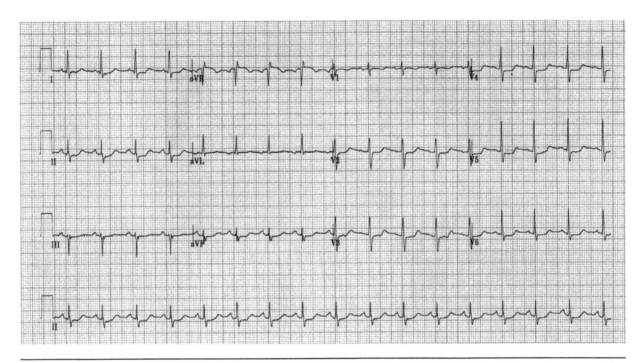

Figure 4.39 A subendocardial, non-ST elevation MI (NSTEMI) demonstrating persistent ST-segment depression.

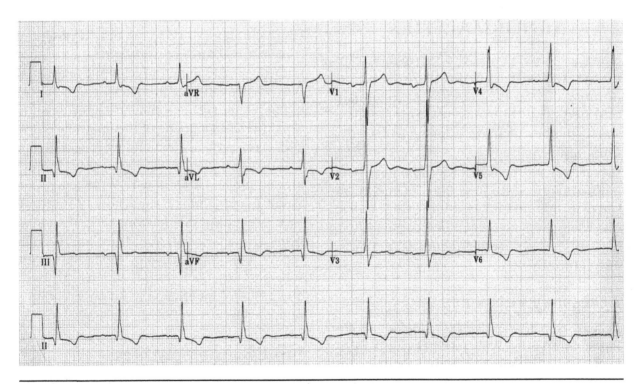

Figure 4.40 A subendocardial, non-ST elevation MI (NSTEMI) demonstrating T-wave inversion in the absence of ST-segment depression.

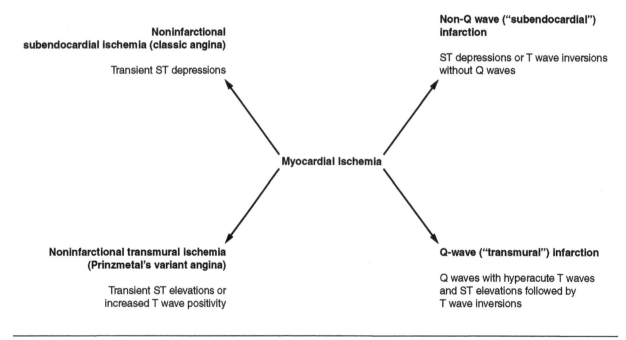

Figure 4.41   Common ECG changes observed in the presence of various forms of myocardial ischemia and infarction.

UNUSUAL ECG ABNORMALITIES

Many less common ECG abnormalities are associated with specific disorders that do not fall into the categories previously outlined. The following sections detail a small number of these unusual ECGs covering abnormalities in electrical signaling (accessory pathway diseases and ion channelopathies), acute infections of the heart (myocarditis and pericarditis), heart muscle diseases (hypertrophic cardiomyopathy), electrolyte disturbances, acute drug effects, and cardiac arrest.

Accessory Pathways

Accessory pathway disorders include Wolff-Parkinson-White (WPW) and Lown-Ganong-Levine (LGL) syndromes. WPW is a common syndrome associated with the preexcitation of the ventricles as a result of the rapid conduction of the atrial action potential through an accessory pathway, bypassing the AV junc-tion, and the associated slowing of the action potential. WPW is characterized by a wide QRS complex (>0.12 s; the action potential travels simultaneously through the accessory pathway and AV junction resulting in a fusion beat) that is reflected in a shortening (<0.12 s) of the PR interval to a similar degree. In addition, there is a slurring of the upstroke of the QRS wave termed the delta wave (see figure 4.42). At first glance WPW resembles bundle branch block. Patients with WPW are prone to atrial arrhythmias, especially PAT and AF (see preceding discussion). PAT develops when the action potential traveling through the AV junction is recycled along the accessory pathway back to the atria, travels back down the AV junction, and so on, creating a rapid atrial rate.

Ion Channelopathies

The ion channelopathies include long QT syndrome, Brugada syndrome, progressive cardiac conduction defect (Lev-Lenegre's syndrome), idiopathic ventricular fibrilla-

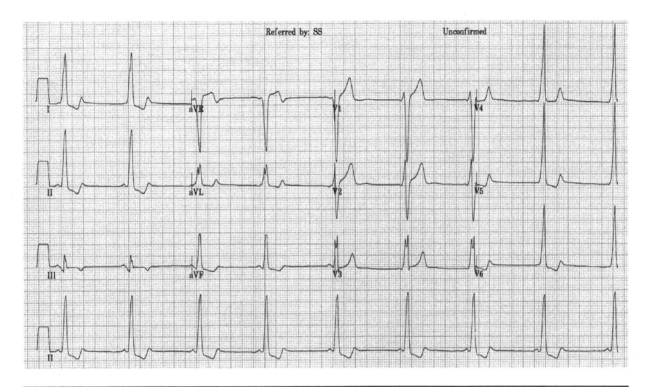

Figure 4.42 WPW showing a characteristically short PR interval and wide QRS complex with a delta wave.

tion (without Brugada ECG changes), and catecholaminergic polymorphic VT (CPVT). This is a rare group of disorders that affect the sodium, potassium, and calcium channels, which are responsible for the regulation of the inflow and outflow of electrical current in the myocardial cells (see chapter 1). The ion channelopathies are genetic disorders that can cause arrhythmia and sudden death. Each ion channelopathy has a subtle, characteristic ECG pattern (see figures 4.43 and 4.44 on p. 90).

Myocarditis

Myocarditis is characterized by an inflammation of the myocardium as a result of a viral or bacterial infection. Myocarditis is often associated with pericarditis; patients present with signs and symptoms suggestive of concurrent myocarditis and pericarditis. Diffuse T-wave inversions with saddle-shaped ST segments are common ECG findings (see figure 4.45 on p. 91).

Acute Pericarditis

Inflammation of the pericardium is commonly caused by viral infections (most commonly implicated are the Coxsackie B. and E. Coli viruses) or bacterial infection, metastatic tumors, collagen vascular disease, myocardial infarction, or uremia. The ECG pattern in pericarditis resembles acute MI except ST-segment elevation is diffuse across all leads (in acute MI, ST-segment elevation is concentrated in anterior or inferior leads) and Q waves are not present (see figure 4.46 on p. 91). Furthermore, acute pericarditis results in atrial repolarization abnormality leading to PR-segment depression in all leads (lead aV_R demonstrates PR-segment elevation).

Hypertrophic Cardiomyopathy (HCM)

HCM is an inherited heart muscle disease that is associated with abnormalities in genes that code for the proteins responsible

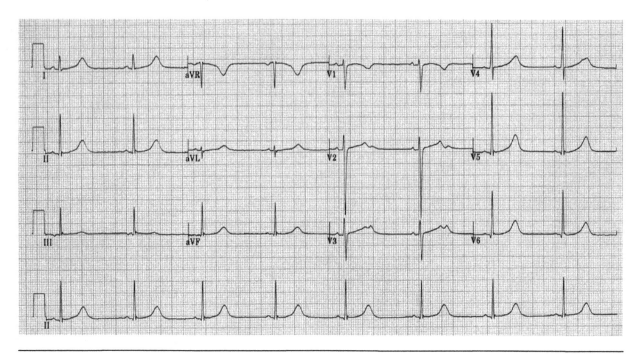

Figure 4.43 Long QT syndrome.

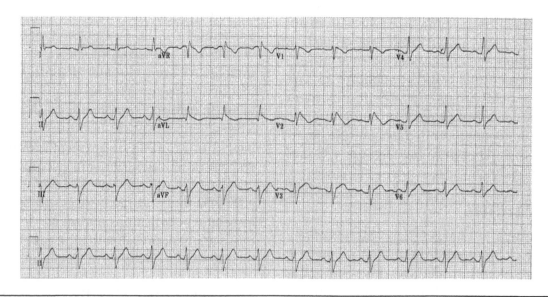

Figure 4.44 Brugada syndrome.

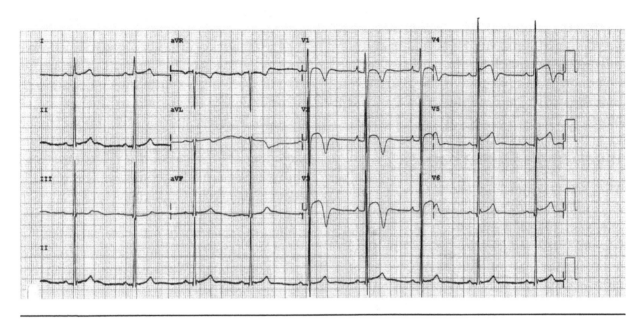

Figure 4.45 Myocarditis.

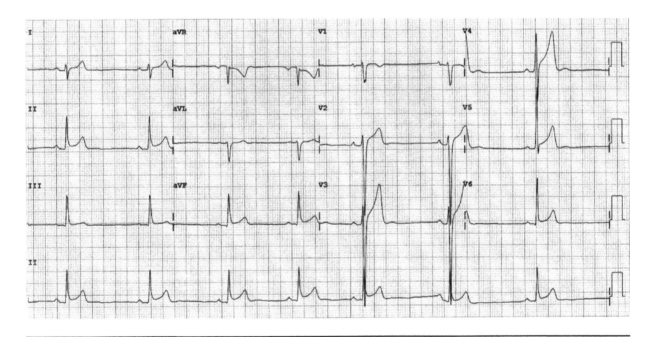

Figure 4.46 Acute pericarditis.

for contraction of the heart (sarcomeric contractile proteins). The ECG is abnormal in 98% of patients with HCM that is characterized by LVH with strain pattern (T-wave inversion in the left precordial leads V_4-V_6) together with pathological Q waves (see figure 4.47).

Electrolyte Disturbances

Abnormal serum concentrations of potassium and calcium can result in profound ECG changes. Hyperkalemia affects both depolarization and repolarization related to serum concentrations. Common changes include narrow, peaking ("tented") T waves with small increases in serum potassium (see figure 4.48) and QRS prolongation resulting in a sine-wave pattern and eventual asystole. In contrast, hypokalemia results in ST-segment depression with prominent U waves and a prolonged QT interval (see figure 4.49).

Acute Drug Effects

A large number of drugs can lead to small, nonspecific changes on the resting ECG (common causes were discussed earlier). People with heart failure and certain arrhythmias use digitalis. Digitalis shortens the repolarization time, which results in a number of effects including shortening of the QT interval and scooping of the ST-T complex. The digitalis effect is different from digitalis toxicity, which can cause complex arrhythmias (see the preceding discussion). In contrast, other antiarrhythmia drugs (e.g., quinidine, procainamide, and disopyramide) cause QT prolongation (see figure 4.50 on p. 94).

Cardiac Arrest

Cardiac arrest is defined as a loss of effective contraction resulting in a loss of cardiac output and the absence of a palpable pulse.

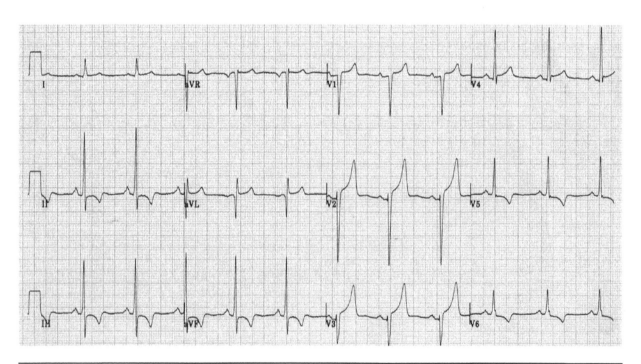

Figure 4.47 Hypertrophic cardiomyopathy (HCM).

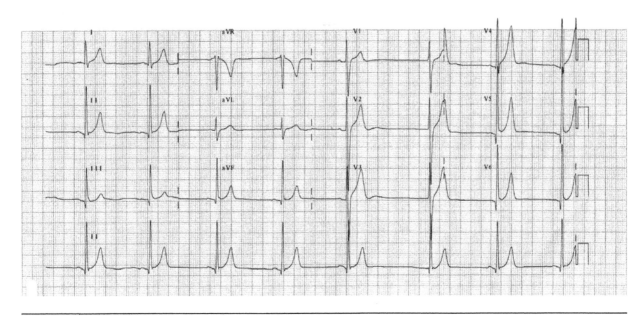

Figure 4.48 Hyperkalemia (peaking T waves).

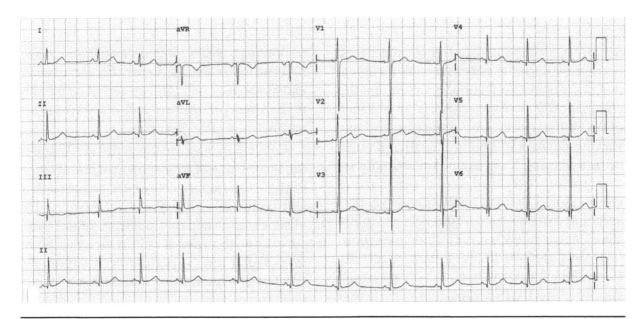

Figure 4.49 Hypokalemia (U waves).

The loss of cardiac output causes blood pressure to fall to zero, which results in a rapid loss of consciousness. Shortly after the heart stops pumping blood, spontaneous breathing ceases (cardiopulmonary arrest). The diagnosis of cardiac arrest can and should be made clinically prior to the connection of an ECG and is identified by an absence of a palpable pulse, absence of heart sounds of auscultation, no blood pressure, cyanosis, and extremity cooling. Pupils may be fixed and dilated (brain hypoxia). Seizure

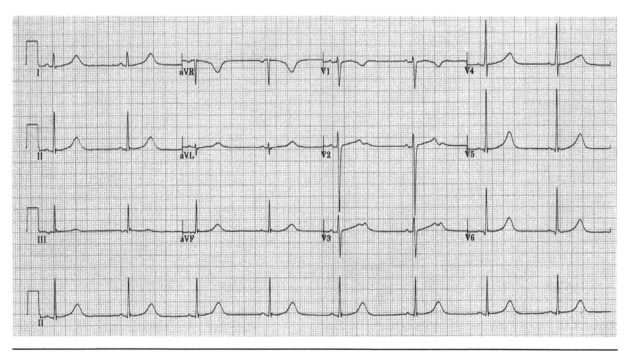

Figure 4.50   QT prolongation caused by quinidine.

activity is a common occurrence. In general, however, all that is necessary for a clinical diagnosis of cardiopulmonary arrest and the commencement of treatment is an absence of breathing. Health care professionals have varying degrees of skill in palpating a pulse or auscultating heart sounds. Furthermore, attempting to record blood pressure is unnecessarily time-consuming, and cyanosis and cold extremities are unreliable and late signs. Accordingly, current guidelines, both for bystander cardiopulmonary resuscitation (CPR) and for in-hospital advanced life support, are that cardiopulmonary arrest is diagnosed on the basis of the absence of detectable breathing.

Cardiac arrest results from the following:

- Ventricular tachyarrhythmias: VF, sustained VT, torsades de pointes, or ventricular flutter.

- Ventricular standstill (asystole): Characterized by a flat line (commonly undulating as a wandering baseline) with occasional junctional or ventricular escape beats. *P-wave asystole* and *ventricular standstill* are terms commonly used to describe the presence of P waves in the absence of a QRS-T complex.

- Pulseless electrical activity (PEA; sometimes termed electromechanical dissociation [EMD]): Characterized by regularly recurring QRS complexes (occasionally P waves) in the absence of a palpable pulse or blood pressure. Common causes of PEA include diffuse myocardial injury, pericardial tamponade, tension pneumothorax, and massive pulmonary embolism.

During resuscitation attempts one or all of these patterns may be observed. Furthermore, cardiac compressions can result in wide sine-wave complexes that should not be confused with intrinsic electrical activity.

This chapter contains information on a wide range of abnormal ECG's. It is not possible to provide key points for the chapter as a whole due to the specific nature of these abnormalities; therefore, we encourage readers to summarize the various abnormalities independently as part of the learning process. However, there are a number of more general key points from the chapter:

▶ On first inspection of an ECG the following should be established to aid interpretation: calibration (standardization make); heart rate; electrical axis; atrial rhythm; ventricular rhythm; AV conduction; presence of unusual complexes.

▶ Bradyarrhythmias (<60 beats·min⁻¹) are unusual rhythms generally associated with a low heart rate and fall under a number of categories including: sinus bradycardia; sinus arrhythmia; AV junctional rhythms; heart block (1st, 2nd [Mobitz type I and type II] and 3rd degree); and left and right bundle branch blocks.

▶ Tachyarrhythmias are rhythms generally associated with elevated heart rates (>100 beats· min⁻¹) and fall under a number of categories including: sinus tachycardia; supraventricular arrhythmias (PAC, PAT, AV junctional rhythms, SVT and AF); ventricular arrhythmias (PVC's, bigeminy, trigeminy, couplets and triplets); ventricular tachycardia (VT, torsade de point, R on T, ventricular fibrillation); and an elevated heart rate without an obvious origin (RVOT-VT and CPVT).

▶ Atrial enlargement is characterized by changes in P-wave duration and morphology, and ventricular enlargement is characterized by changes in the QRS complex, ST segment, and T-wave duration and morphology.

▶ Myocardial ischemia results in ST segment depression that is transient in nature. Myocardial infarction results in QRS and ST-T segment changes in various stages reflecting the progression of the condition.

▶ A large number of less common conditions exist that result in an abnormal resting ECG. Understanding the structure and function of the heart, the electrical and contractile properties of the heart, and the normal ECG will assist in identifying the etiology (causes) of abnormal findings in a variety of diseased states.

5

ECG During Exercise

Exercise is commonly used in the clinical setting to create a potent physiological stress to the cardiovascular system. Exertion-related symptoms are the primary indication for exercise testing. Examining the cardiopulmonary responses to exercise helps practitioners identify underlying cardiovascular disease and elucidate the mechanisms underlying abnormal responses. As exercise intensity (load) increases, the following responses are normally observed (chapter 1 describes the cardiovascular response to exercise in detail):

- Increased heart rate (HR)
- Increased mean arterial blood pressure (BP) due to an increased systolic BP (SBP) and relatively stable diastolic BP (DBP)
- Increased stroke volume (SV)
- Increased cardiac output increases (Q = HR $\times$ SV)
- Increased myocardial oxygen demand ($M\dot{V}O_2$ = HR $\times$ SBP)

The increased myocardial oxygen demand, combined with an altered sympathovagal balance and fluctuating hormonal milieu, acts as a significant stressor to the heart during and immediately following exercise. Exercise stress testing may therefore be employed to identify people with compromised myocardial blood flow or an underlying pathology predisposing them to exercise- or exertion-related arrhythmias, or both. Exercise stress testing may also be used to establish the functional capacity of a patient with cardiovascular disease and evaluate the efficacy of pharmacotherapy. Exercise stress testing is therefore valuable in the following settings:

- Functional testing: Submaximal and maximal exercise testing can establish functional and cardiorespiratory capacity.

- Diagnostic testing: Submaximal and maximal exercise testing can be used to investigate people presenting with symptoms associated with coronary artery disease (e.g., angina, dyspnea, ECG abnormalities, exercise- or exertion-associated collapse or syncope, possible history of a cardiac event).

- Disease severity and prognostic testing: Submaximal exercise testing can indicate disease severity and prognosis.

- Pre- and postdischarge testing: Submaximal and maximal (with care) exercise testing can be used to evaluate a patient prior to discharge following a myocardial infarction (MI) to establish functional capacity and the efficacy of drug or treatment therapy. Furthermore, submaximal and maximal testing is valuable in the post-discharge assessment of patients following MI, percutaneous transluminal angioplasty (PCTA), and coronary artery bypass graft surgery (CABG). It can be used to assess exercise capacity, prescribe exercise, establish drug efficacy, and identify the requirement for further intervention.

Initial assessment of a patient is usually indicated in the following situations:

- Symptoms indicating myocardial ischemia (e.g., chest tightness, dyspnea)
- Exertional syncope
- Exertion- or exercise-related palpitations or arrhythmias
- Cardiomyopathy (dilated and hypertrophic)
- Flow-limiting valve lesions when considering surgery or transplant
- Differential diagnosis of exercise limitation
- Suspected or known pulmonary hypertension

Integrated cardiopulmonary exercise stress testing, sometimes termed CPXT or CPEX, involves the simultaneous evaluation of cardiovascular response (i.e., ECG and blood pressure) and pulmonary response (i.e., ventilation and gas exchange). Such testing is valuable when examining people who may have both cardiac and pulmonary limitations (i.e., points 5 and 6 above). Although the cost and expertise required to run cardiopulmonary exercise stress testing often preclude its use from general clinical practice, it is a valuable addition to exercise stress testing for all patient populations and should be employed whenever possible. A detailed discussion of cardiopulmonary exercise testing is beyond the scope of this text and is covered in detail elsewhere.* This chapter focuses on identifying obstructive coronary artery disease resulting in exercise-induced myocardial ischemia and exertion- and exercise-related arrhythmias.

NORMAL ECG RESPONSES DURING AND POSTEXERCISE

The physiological stress of exercise elicits a predictable cascade of responses on the ECG. Analysis of the ECG during and postexercise should always include the following:

- Heart rate and the relationship with exercise intensity
- Heart rhythm
- Measurements and morphology including QRS complexes, ST-segment changes, and QT interval

The Heart Rate and Workload Relationship

A consistent and reproducible linear relationship exists between heart rate (HR) and workload (heart rate increases linearly with exercise intensity up to the maximum heart rate, or HR_{max}). By plotting heart rate against workload, the practitioner can observe the chronotropic response of the heart. A sudden increase in heart rate during or postexercise may indicate the development of a tachyarrhythmia. In contrast, a rapid fall in heart rate immediately following exercise, together with a dramatic fall in blood pressure, may precipitate presyncope or syncope associated with a vasovagal response. A failure of heart rate to rise or an abnormally slow increase during exercise (termed chronotropic incompetence) may indicate electrical conduction pathway disease. An abnormal heart rate response during recovery is a strong predictor of all-cause mortality in a

---

*For further reading see:

1. Bibbons, R.J., Balady, G.J., Bricker, J.T., Chaitman, B.R., Fletcher, G.F., Froelicher, V.F., Mark, D.B., McCallister, B.D., Mooss, A.N., O'Reilly, M.G. and Winters, W.L. Jnr. ACC/AHA 2002 Guideline Update for Exercise Testing: A report of the ACC/AHA Task Force on Practice Guidelines (Committee on Exercise Testing). *2002. American College of Cardiology Web site. Available at: www.acc.org/*

2. *ACSM's Guidelines for Exercise Testing and Prescription 5th Ed.* Williams & Wilkins, Media, PA, USA.

clinical population (meaning they have been referred). (Apparent chronotropic incompetence may also be drug induced—for example, beta-blockers or non-dihydropyridine calcium channel antagonists).

The increase in heart rate during exercise causes a shortening of diastole and systole. Despite these changes, sinus rhythm is maintained in the normal heart. Recognition of abnormal rhythms is important in identifying brady- and tachyarrhythmias during exercise.

A number of expected ECG changes occur during exercise in the normal heart (see the section, Expected ECG Changes in the Normal Heart). Differentiating normal from abnormal changes assists in the diagnosis of underlying cardiovascular disease.

Rhythm Recognition

Exercise-induced arrhythmias can be determined by answering similar questions to those posed when evaluating an ECG at rest:

- What is the atrial rhythm? In the normal heart, P waves should be clearly identifiable because they occur prior to every QRS complex. Identification of a P wave may be difficult at high heart rates during exercise because the P wave may be superimposed on the T wave of consecutive beats.

- What is the ventricular rhythm? In the normal heart the duration of the PR interval and the QRS complex are shortened.

- Is the AV conduction normal? QRS complexes should always be preceded by a fixed PR interval.

- Are there any unusual complexes? The morphology of some ECG parameters may change throughout exercise in a predetermined fashion (discussed later); however, P waves, QRS complexes, and T waves will have the same nomenclature in a single lead at any given time.

- Is the rhythm dangerous? Dangerous rhythms that indicate the need to stop exercise immediately during stress testing include ventricular fibrillation (VF), sustained ventricular tachycardia (VT), and ST-segment elevation ($\geq$1 mm) in leads without diagnostic Q waves (other than V_1 or aV_R). Relative ECG stop test indicators include ST or QRS changes such as excessive ST-segment depression (>2 mm of horizontal or downsloping ST-segment depression) or marked axis shift; arrhythmias other than VT including multifocal PVCs, triplets, supraventricular tachycardia (SVT), heart block, and bradyarrhythmias; and the development of bundle branch block (see table 5.1 on p. 100 for a full listing of exercise stress test stop test indicators).

Expected ECG Changes in the Normal Heart

The altered action potential duration, conduction velocity, and contractile velocity associated with the increase in heart rate during exercise results in a number of ECG changes in normal people, including the following:

- RR interval decreases
- P-wave amplitude and morphology undergo minor changes
- Septal Q-wave amplitude increases
- R-wave height increases from rest to submaximal exercise and then reduces to a minimum at maximal exercise
- The QRS complex experiences minimal shortening
- J-point depression occurs
- Tall, peaked T waves occur (high interindividual variability)
- ST segment becomes upsloping
- QT interval experiences a rate-related shortening (see table 5.2 on p. 100)
- Superimposition of P waves and T waves on successive beats may be observed

The depression of the J point in normal people result in marked ST-segment upsloping

Table 5.1 Exercise Stress Test Stop Test Indicators

Absolute stop test indicators
Drop in SBP >10 mmHg from baseline despite increase in workload and when accompanied by other evidence of ischemia
Moderate to severe angina
Increasing nervous system symptoms (e.g., ataxia, dizziness, presyncope)
Signs of poor perfusion (cyanosis or pallor)
Sustained VT
ST-segment elevation (≥1 mm) in leads without diagnostic Q waves (other than V_1 or aV_R).
Technical difficulties in monitoring ECG and BP
Subject requests to stop

Relative stop test indicators
Drop in SBP (≥10 mmHg from baseline BP despite an increase in workload in the absence of other evidence of ischemia)
Fatigue, shortness of breath, wheezing, leg cramps, or claudication
Increasing chest pain
Hypertensive response (SBP >250 mmHg and/or DBP >115 mmHg)
ST or QRS changes such as excessive ST-segment depression (>2 mm of horizontal or downsloping ST-segment depression) or marked axis shift
Arrhythmias other than VT including multifocal PVCs, triplets, supraventricular tachycardia (SVT), heart block, and bradyarrhythmias
Development of bundle branch block or an interventricular conduction delay that cannot be distinguished from sustained VT

This article was published in *Journal of the American College of Cardiology*, R.J. Gibbons et al., "Guideline update for exercise testing: A report of the ACC/AHA task force on practice guidelines (Committee on Exercise Testing)," pg. 6 Copyright Elsevier 2002.

Table 5.2 Normal Relationship Between Heart Rate and QT Interval During Exercise

Heart rate	QT interval(s)
60	0.33-0.43
80	0.29-0.38
100	0.27-0.35
120	0.25-0.32
140	0.23-0.28
160	0.21-0.26
180	0.19-0.24

QT interval is best measured in V_3 for exercise test data because it usually presents the largest T-wave amplitude (at rest, lead II is commonly used).

associated with competition between normal repolarization and delayed terminal depolarization forces, rather than ischemia. The J-point depression and tall, peaked T waves observed during exercise may be sustained during recovery in normal people.

EVALUATION OF PEOPLE WITH KNOWN OR SUSPECTED EXERCISE- OR EXERTION-RELATED ARRHYTHMIAS

An increased sympathetic drive combined with vagal withdrawal and changes in extracellular and intracellular electrolytes, pH, and oxygen tension have a profound impact on conduction tissue automaticity and reentry. For this reason, exercise acts as a potent arrhythmogenic stimulus. Exercise- and exertion-induced arrhythmias can occur in diseased and healthy populations and are categorized as supraventricular arrhythmias, ventricular arrhythmias, and sinus node dysfunction. People with known or suspected exercise- or exertion-related arrhythmias present with syncope, presyncope, symptoms of rapid heart rates or palpitations, dyspnea, reduced exercise performance, or malaise.

Ventricular Arrhythmias

Exercise- or exertion-induced tachycardias may reflect the presence of ischemia, structural abnormalities that induce an abnormal cardiac response, or increased circulating catecholamines. Isolated premature ventricular ectopics (VEs) occur in approximately 30 to 40% of healthy people during exercise. Those presenting with isolated premature VEs should be reassured of the benign nature and lack of clinical significance. In coronary artery disease patients the prevalence is higher (50 to 60%). Exercise may lead to the disappearance of VEs in some people presenting with VEs at rest. The clinical significance of this suppression is not fully understood, although it may be associated with significant coronary artery disease.

The value of exercise stress testing in those with suspected exercise- or exertion-induced sustained ventricular tachycardia (VT) is variable with poor reproducibility because of the difficulty in uncovering the arrhythmia during exercise. In contrast, exercise stress testing is valuable when examining the link between arrhythmia and exertional symptoms and when evaluating antiarrhythmic drug therapy. Furthermore, VT is usually confined to patients with structural or electrical pathologies including cardiomyopathy, ion channelopathies (long QT and Brugada syndromes), valvular heart disease, severe ischemia, or a family history of sudden death (see figure 5.1 on p. 102). Exercise-induced VT may also be present with other syndromes including right ventricular outflow tract VT (RVOT-VT) in the normal heart (see chapter 7).

Despite these shortcomings, ECG exercise stress testing may provide the necessary substrate for arrhythmogenesis and act as a valuable preliminary investigation prior to an electrophysiological study. Furthermore, exercise-induced VT may be an important prognostic indicator given that a greater mortality rate is observed in those presenting with complex exercise-induced arrhythmia. In patients undergoing antiarrhythmic therapy, exercise-induced VT is associated with a high risk of sudden death.

Nonsustained VT is also uncommon and only of clinical importance in the previously discussed patient groups.

An increasing frequency of VEs, multifocal VEs, coupling, and salvos are stop test indicators. A full cardiovascular follow-up is indicated to establish the underlying causes.

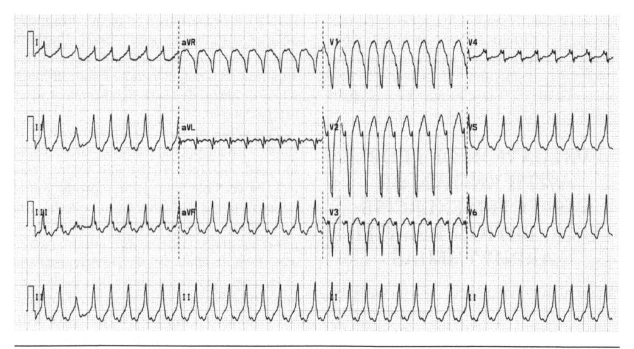

Figure 5.1   Exercise-induced VT in a patient presenting with hypertrophic cardiomyopathy (HCM) with a ventricular rate of 197 beats·min^{-1}.

Supraventricular Arrhythmias

Isolated premature atrial contractions (PACs) are common and of no clinical significance. In contrast, an exercise-induced sustained supraventricular tachycardia (SVT) likely associated with atrial flutter or atrial fibrillation (AF) is of greater clinical significance and may occur as a result of organic heart disease or endocrine, metabolic, or drug effects including hyperthyroidism, alcohol consumption, cocaine use, and digoxin toxicity. Figure 5.2 demonstrates an exercise-induced SVT as a result of excessive alcohol consumption prior to maximal exercise stress testing (see case study 1 in chapter 7).

Exercise-induced SVT is characterized by a rapid heart rate that often increases abruptly leading to a loss of linearity in the heart rate/workload relationship. The QRS duration is usually normal with an irregular RR interval and a loss of association between P waves and QRS complexes. At high ventricular rates, SVT may lead to LBBB that resembles VT (see figure 5.3). The practitioner must take care to avoid misdiagnosis of VT in this case.

The ventricular response to SVT is controlled by the AV node (see the later discussion of people with accessory pathways). Accordingly, the rate of repolarization and effective refractory period of the AV node governs the conduction rate. Antiarrhythmic drugs used for rate control in patients with AF target the conduction rate of the AV node. Rate control at rest does not indicate rate control at higher heart rates, and exercise stress testing may be important in unmasking rate-dependent AF and prescribing appropriate medication.

In addition to identifying SVT, ECG exercise stress testing may be used to evaluate the risk of developing rapid ventricular response during atrial arrhythmia (e.g., in those with accessory pathway disease). The abrupt loss of preexcitation in those with WPW syndrome suggests a longer antegrade

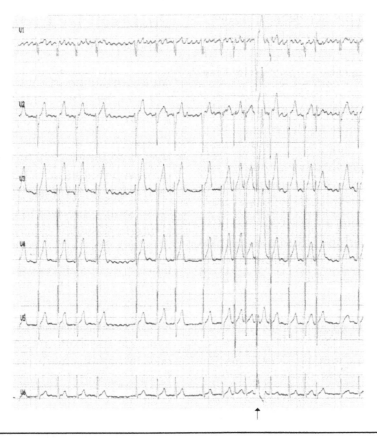

Figure 5.2 Exercise-induced SVT (AF) as a result of excessive alcohol consumption prior to exercise stress testing. Notice the characteristic rapid AF with variable ventricular conduction (ventricular ectopic, indicated by an arrow).

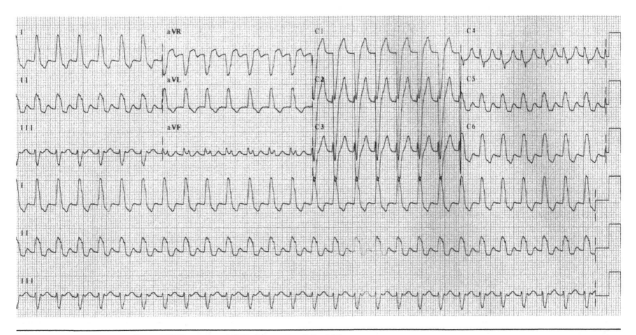

Figure 5.3 Exercise-induced SVT causing LBBB that resembles VT.

refractory period in the accessory pathway than in the AV node. It is therefore unlikely that rapid ventricular rates will occur at heart rates above the rate leading to the loss of preexcitation.

In some cases the exercise-induced SVT may be sustained following the cessation of exercise and despite vagal reflex maneuvers (e.g., carotid massage, Valsalva maneuver). If the patient is symptomatic and compromised with falling blood pressure, electrical or pharmacologic cardioversion may be required (see case study 1 in chapter 7). Exercise stress testing may be valuable in the follow-up care of the patient to establish the efficacy of drug therapy.

Sinus Node Dysfunction

Exercise stress testing may be used to distinguish people with a resting bradycardia and normal chronotropic response to exercise (i.e., athletes) from those with resting bradycardia and an inappropriate chronotropic response to exercise. Chronotropic incompetence is commonly defined as a failure to achieve 80 to 85% of age-predicted maximum heart rate in the absence of a noncardiopulmonary exercise limitation. The use of exercise stress testing for the identification of sinus node dysfunction is poorly predictive and therefore limited.

Postexercise Chronotropic Response

An abnormal heart rate recovery (defined as a change from the peak exercise heart rate to a heart rate measured two minutes postexercise of ≤12 beats·min^{-1}) is strongly predictive of an increase in all-cause mortality in a clinical population. In contrast, a rapid fall in heart rate immediately following exercise is commonly observed in people with exercise-related vasovagal syncope.

Syncope is the transient loss of consciousness and postural tone and rarely occurs during exercise, but when it does, it is often an ominous sign of structural heart disease. In contrast, however, recurrent idiopathic syncope has been described following exercise in patients with normal hearts and has been reported to be particularly prevalent in endurance-trained athletes (see figure 5.4). The pathophysiology of exercise-induced vasovagal syncope is not fully understood but is associated with an abnormal chronotropic and vasoconstrictor response to orthostatic challenge following exercise. The dramatic fall in cardiac output and blood pressure results in cerebral hypoperfusion and syncope. Exercise-induced vasovagal syncope is common and, in the presence of a negative cardiovascular workup, is benign. A number of simple inter-

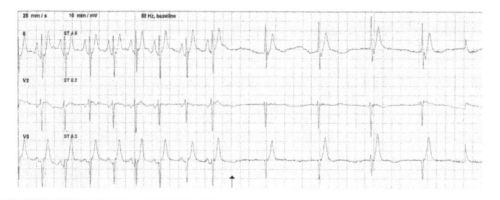

Figure 5.4 An ECG of an athlete suffering from postexercise vasovagal syncope. Note the rapid, instantaneous fall in heart rate following the cessation of exercise (arrow).

ventions can be employed to avoid syncope, including continuation of exercise and leg crossing (see case study 2 in chapter 7).

ECG EXERCISE STRESS TESTING IN THE DIAGNOSIS OF CORONARY ARTERY DISEASE (CAD)

The vast majority of ECG exercise stress tests are performed in adults with known or suspected CAD. During exercise, the myocardial oxygen demand is increased in relation to heart rate and contractility, which are both closely related to exercise intensity (workload). Limitations in coronary blood flow are closely related to myocardial oxygen demand. Accordingly, a limitation in coronary blood flow will be uncovered during exercise when the threshold between oxygen demand and oxygen supply is reached (myocardial ischemia occurs above that threshold). This threshold varies depending on the degree of a person's blood flow limitation. Furthermore, the location of the exercise-induced ischemia varies depending on the location of the limita-

tion to blood flow (i.e., which coronary artery or arteries are affected). Ischemia manifests in the following ways:

- The patient may report symptoms of myocardial ischemia (angina pectoris, dyspnea).
- ST-segment changes may be present on the ECG.
- Cardiac function may be impaired resulting in a fall in stroke volume and a reduction in functional capacity.

ST-Segment Changes

ST-segment changes are widely recognized as the primary indicators of myocardial ischemia during exercise. Three main types of ST-segment changes are associated with exercise-induced myocardial ischemia (see figure 5.5):

- *ST-segment elevation:* On an otherwise normal ECG, exercise-induced ST-segment elevation indicates transmural ischemia (caused by spasm or critical lesion) that is highly arrhythmogenic. ST-segment elevation is common following a Q-wave infarction but

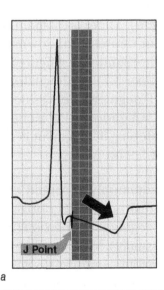
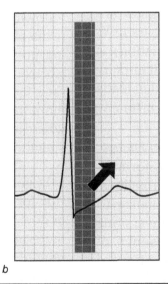
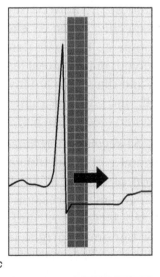

a   b   c

Figure 5.5   A schematic representation of variability of exercise induced ST-segment depression indicative of myocardial ischemia (the j-point occurs 0.04s after the end of the QRS complex). Upsloping ST-segment depression may be a normal variant. *(a)* Downsloping, *(b)* upsloping, and *(c)* horizontal ST-segment depression.

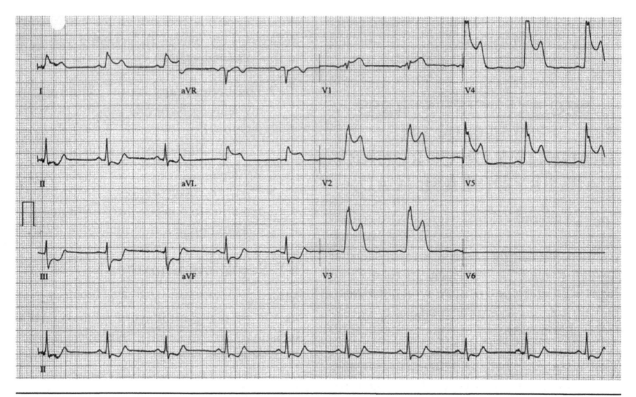

Figure 5.6   Transmural myocardial infarction demonstrating ST-segment elevation in the anterior and lateral leads and reciprocal ST-segment depression in the inferior leads, typical of anterior MI.

is very rare in people with normal resting ECGs. The location of the ischemia can be identified from the leads in which ST-segment change is observed with the exception of V_1 and aV_R, where ST-segment changes are nonspecific. When ST elevation occurs in V_2 through V_4, the LAD (left axis deviation) is involved; in the lateral leads, the left circumflex and diagonals are involved; and in leads II, III, and aV_F, the right coronary artery is involved (see figure 5.6).

• *ST-segment normalization or no change:* Ischemia may be indicated by a normalization of resting ST-segment abnormalities. ST-segment depression may normalize to the isoelectric line, and T-wave inversions may become upright in the presence of ischemia. This pseudo-normalization of the ST segment should be viewed as ST-segment elevation and indicative of myocardial ischemia.

• *ST segment depression:* The most commonly observed ST segment change as a result of exercise induced ischemia is depression of the J point and the ST slope over the following 0.06s (3mm). There are 3 main patterns of ST segment depression: downsloping, upsloping and horizontal (see figure 5.5). Horizontal (see figure 5.7) or downsloping (see figure 5.8) ST segment depression is more indicative of subendocardial ischemia where a slowly upsloping ST segment is nonspecific (see figure 5.9 on p. 108).

In general, the commonly accepted criterion for a positive test is 1 mm of horizontal or downsloping ST-segment depression (zero or negative slope visually). Most exercise labs require 1 mm of horizontal or downsloping ST-segment depression in two or more contiguous leads. Unlike Q waves on the resting ECG, ST depression does not identify the location

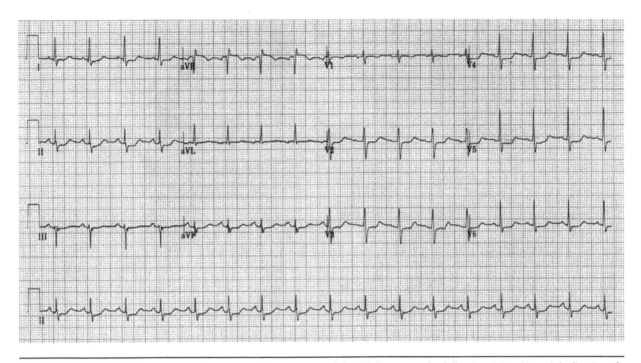

Figure 5.7 Horizontal ST-segment depression of the J point and the ST slope over the following 0.06 s (3 mm), indicative of subendocardial ischemia.

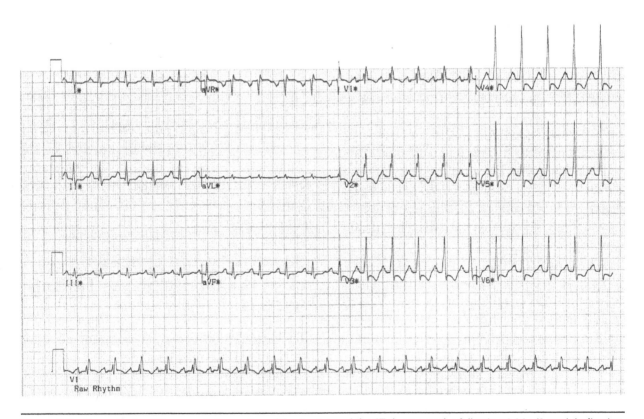

Figure 5.8 Downsloping ST-segment depression of the J point and the ST slope over the following 0.06 s (3 mm), indicative of subendocardial ischemia.

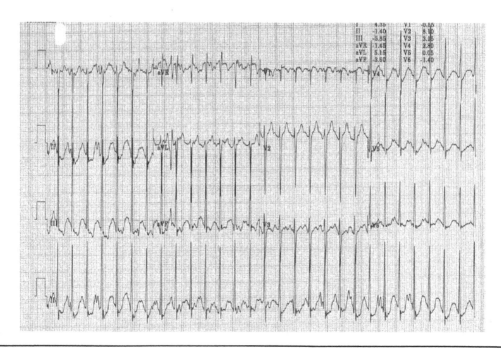

Figure 5.9   Slowly upsloping ST-segment depression commonly observed in normal people and nonspecific for myocardial ischemia.

of myocardial ischemia. However, the greater the number of leads showing ST-segment depression, the more severe the disease.

ST-segment changes following exercise may be valuable in the assessment of the clinical significance of ST-segment changes. Immediately following exercise, heart rate and blood pressure fall resulting in a rapid reduction in myocardial oxygen consumption. The fall in oxygen demand can result in a normalization of the ST segment. However, more commonly, ST segment remains elevated. A significant ST-segment depression occurring solely postexercise may indicate myocardial ischemia and require follow-up.

When interpreting ST segment, lead V_5 alone consistently outperforms the inferior leads and the combination of lead V_5 with lead II. In a person with a normal resting ECG and an absence of prior myocardial infarction, the chest leads alone are a reliable marker of CAD (monitoring limb leads adds little to the diagnostic yield). Exercise-induced ST-segment depression solely in the inferior leads is of limited clinical significance in those with normal resting ECGs.

Computer Processing

Computer-derived summaries of ST-segment measurements are available from most exercise stress-testing systems (see figure 5.10). As with resting ECG analysis, care is warranted in the use of these derived averages; ECG recordings of the raw (unprocessed) ECG at baseline, throughout exercise, and during recovery should be provided alongside the summary. At present, none of the computer-generated scores or measurements have been validated sufficiently to recommend their widespread use.

Sensitivity and Specificity of ECG Exercise Stress Testing in CAD

The interpretation of ST-segment changes during exercise may be affected by a number of factors (see table 5.3). Exercise stress testing to identify myocardial ischemia should be avoided in people presenting with left bundle branch block (LBBB) at rest. Exercise-induced ST-segment depression in the anterior chest

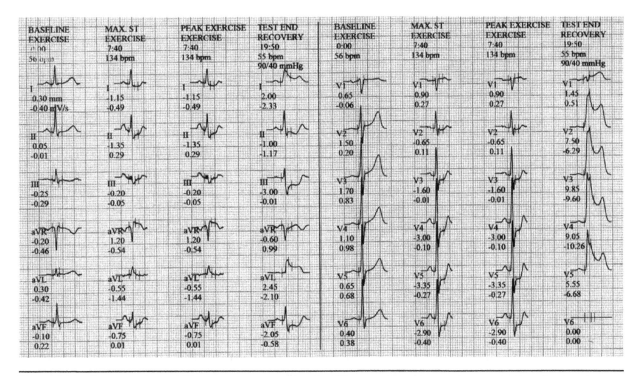

Figure 5.10 Computer-derived ST-segment summary showing ST-segment measurement averages at baseline, maximum exercise, peak exercise, and test end/recovery for limb and precordial leads.

Table 5.3 Causes of False Negative and Positive Exercise Stress Test Results

Causes of false negative	Causes of false positive
Failure to reach ischemic threshold as a result of functional limitation (physical incapacity or drug effect) or early test termination	Repolarization abnormalities at rest (e.g., LBBB, RBBB)
Failure to identify non-ECG signs and symptoms	Cardiac hypertrophy
Angiographically significant disease compensated by collateral circulation	Accelerated conduction defects (e.g., WPW syndrome)
Inadequate number of leads to identify ECG changes	Nonischemic cardiomyopathy
Repolarization abnormalities at rest (e.g., LBBB, RBBB)	Hypokalemia
Technical or observer error	Vasoregulatory abnormalities
	Mitral valve prolapsed
	Pericardial disorders
	Coronary spasm in the absence of significant CAD
	Anemia
	Female gender
	Technical or observer error

Adapted from *ACSM's guidelines for exercise testing and prescription*, 8th ed. (Philadelphia, PA: Williams & Wilkins), 147.

leads (V_1, V_2, V_3) should not be used to diagnose ischemia in people with right bundle branch block (RBBB). In contrast, ST-segment depression in the left chest leads (V_4, V_5, V_6) may indicate myocardial ischemia. However, the appearance of LBBB during exercise, termed rate-dependent LBBB, may indicate myocardial ischemia and require follow-up.

The diagnosis of CAD in women presents difficulties compared to men because of differences in body habits, exercise physiology, coronary physiology, and the prevalence of CAD. Although cardiovascular disease is one of the primary causes of death in women, exceeding the mortality of breast cancer by a factor of 11, the prevalence of CAD in women is lower than that in men. The ST-segment response to exercise appears to be gender related from an early age and, combined with the lower prevalence of CAD, results in difficulties in interpreting the ECG during exercise stress testing.

Exercise-induced ST-segment depression is less sensitive in women reflecting a lower prevalence of severe CAD and a reduced ability to attain maximum exercise, or aerobic, capacity. The lower specificity of ST-segment depression has been explained by lower disease prevalence, greater prevalence of mitral valve prolapse and syndrome X, differences in microvascular function, and hormonal differences. Using non-ECG end points including exercise capacity, heart rate and blood pressure response, and the presence of cardiac symptoms, all improve the sensitivity and specificity of ECG exercise stress testing in women. Furthermore, the stress testing limitations in females have translated into clinical pretest probability guidelines that are gender specific.

The sensitivity (percentage of people with CAD who have abnormal tests) and specificity (the percentage of normal people free of CAD that have normal tests) of ECG exercise stress testing for the detection of CAD is highly variable and depends on a number of factors. The sensitivity of ECG exercise stress testing is negatively affected by false negative results (no or nondiagnostic ECG changes in patients with CAD), which are caused by a number of factors (see table 5.3). False positive tests (ECG changes indicative of CAD in people without disease) have a detrimental effect on the specificity of ECG exercise stress testing (see table 5.3). The sensitivity and specificity of ECG exercise stress tests vary dramatically and can be extremely low if care is not taken to avoid the limiting factors listed in table 5.3. Sensitivity is improved with patients with a greater severity of disease (e.g., when a greater number of coronary vessels are involved or there is a greater degree of occlusion). The inclusion of data examining the duration of exercise (achieved METs), SBP response, HR_{max}, rate pressure product, and symptoms of angina or dyspnea all improve the sensitivity and specificity of the ECG exercise stress test. In general, sensitivity and specificity of ECG exercise stress testing ranges from 60 to 80%. A full discussion of the characteristics of ECG exercise stress test performance is beyond the scope of this book and is covered in detail elsewhere (see ACSM, 2000; Gibbons et al., 2002).

▶ Exertion-related symptoms are the primary indication for exercise testing; however, it is also valuable in the assessment of functional and cardiorespiratory capacity, disease severity and prognosis, and pre- and postdischarge testing.

▶ Cardiopulmonary exercise stress testing is a valuable addition to exercise stress testing for all patient populations and should be employed whenever possible.

▶ Expected ECG changes during exercise include a decreased RR interval, minor changes in P-wave amplitude and morphology, an increase in septal Q-wave amplitude, and R-wave height increases from rest to submaximal exercise that then reduce to a minimum at maximal exercise.

▶ Minimal shortening of the QRS complex; J-point depression; tall, peaked T waves (high interindividual variability); upsloping ST segment; rate-related shortening of the QT interval; and superimposition of P waves and T waves on successive beats may be observed in normal, healthy individuals.

▶ Exercise-induced arrhythmias can be determined by answering similar questions to those posed when evaluating an ECG at rest: What is the atrial rhythm? What is the ventricular rhythm? Is the AV conduction normal? Are there any unusual complexes? Is the rhythm dangerous?

▶ The vast majority of ECG exercise stress tests are performed in adults with known or suspected CAD. The three main types of ST-segment changes associated with exercise-induced myocardial ischemia are ST-segment elevation (transmural ischemia), ST-segment normalization or no change (pseudo-normalization), and ST-segment depression (subendocardial ischemia).

THE ATHLETE'S HEART

Part III focuses specifically on the athlete's heart. Chapter 6 examines the physiologic adaptations associated with chronic physical training manifest as cardiac enlargement and increased thickness of the walls, additional heart sounds on auscultation, sinus bradycardia, and ECG anomalies. These physiologic changes may mimic those observed in pathologic processes, so accurate diagnosis is required for differentiating a physiologic from a pathologic substrate. This is of particular importance for those pathologies that cause sudden cardiac death in athletes. An understanding of the anomalies often present in athletic individuals at rest and during exercise is especially important for those working in sporting environments. Chapter 7 then presents a number of case studies to highlight the problems encountered in dealing with athletes' hearts.

6

The Athlete's Heart: Physiological Versus Pathological

Regular physical training is associated with a number of unique structural and functional adaptations that together are termed *the athlete's heart (AH)*. Hypertrophy of the ventricles, an increase in cardiac chamber size, and enhanced ventricular filling in diastole result in an increased stroke volume (and cardiac output) at rest and throughout exercise. The augmented blood flow at rest observed in athletes is responsible for the characteristic third and fourth heart sounds on auscultation (see chapter 1). A low resting heart rate, often below 60 beats·min⁻¹ (bradycardia), combined with a lower heart rate at a given submaximal workload and lower maximum heart rate (Whyte, George, et al., 2008; see figure 6.1) compared with sedentary controls are further characteristics of AH. The alterations in resting heart rate and the chronotropic response of the heart during

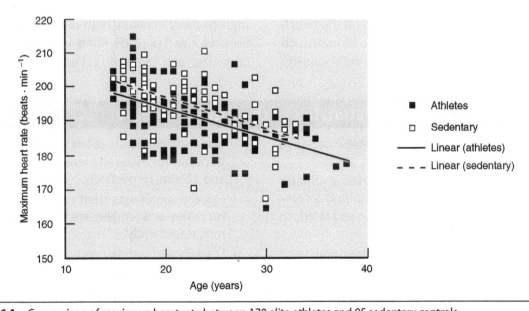

Figure 6.1   Comparison of maximum heart rate between 170 elite athletes and 95 sedentary controls.

Adapted, by permission, from G. Whyte et al., 2008, "Training induced changes in maximal heart rate," *International Journal of Sports Medicine* 29(2): 129-133.

exercise are not fully understood; however, an increased vagal tone possibly associated with the functional effect of an increased stroke volume may be responsible. The altered sympathovagal control of the heart combined with structural changes are likely responsible for an array of ECG anomalies that are observed in athletes.

A large heart and resting bradycardia are well-recognized features of AH and manifest as an increased cardiothoracic ratio on plain chest radiographs, a displaced and forceful cardiac apical impulse on physical examination, additional heart sounds, and ECG anomalies. These physiological adaptations may mimic pathology (Sharma et al., 1997), and differentiating pathological from physiological changes of the heart is important given the former's association with sudden cardiac death in young athletes.

CARDIAC STRUCTURE AND FUNCTION IN HIGHLY TRAINED ATHLETES

Training-induced adaptations in cardiac structure have been traditionally divided into two main types depending on the specific nature of the hemodynamic load placed on the heart. This supposition is based on the Morganroth hypothesis, which suggests a relationship between cardiac adaptation and the specific hemodynamic load imposed by the endurance or strength exercise (Naylor et al., 2008). Endurance training imposes an increased preload on the heart that is associated with a sustained increase in venous return combined with a mild to moderate increase in blood pressure resulting in an increased afterload. As a result of this increased preload and afterload, left ventricular cavity dilatation with minor increases in wall thickness occurs. Resistance training, in contrast, leads to dramatic elevations in blood pressure as a result of simultaneous muscular contraction and Valsalva maneuver. This transiently increased afterload results in an increased left ventricular wall thickness and a relatively normal-sized left ventricular cavity dimension (see figure 6.2). Training rarely conforms to this rigid dichotomy of cardiovascular responses, however; most programs combine endurance and strength training. Thus, the literature supporting a simple division of endurance and strength athletes does not exist.

In the vast majority of athletes, the cardiac adaptations exhibited tend to reflect a combination of responses to increased preload and afterload. Indeed, the largest athletic hearts are observed in athletes who combine elements of strength and endurance exercise during training, often undertaking high-resistance exercise for prolonged periods, such as rowing, cycling, canoeing, and swimming (Pelliccia et al., 1991;

Characteristics of the Athlete's Heart

- Training results in a reversible increase in left ventricular muscle mass (LVM).
- The largest hearts are observed in athletes participating in events requiring a combination of both endurance and strength or power.
- Adaptations are similar between males and females and juniors (<21 years) and seniors; however, males exhibit larger hearts than females, and seniors have larger hearts than juniors.

- Upper normal limits of left ventricular wall thickness for males and females are 14 mm and 12 mm, respectively.
- Upper normal limits of left ventricular cavity for males and females are 66 mm and 60 mm, respectively.
- Diastolic and systolic function are normal or enhanced in the athlete's heart.

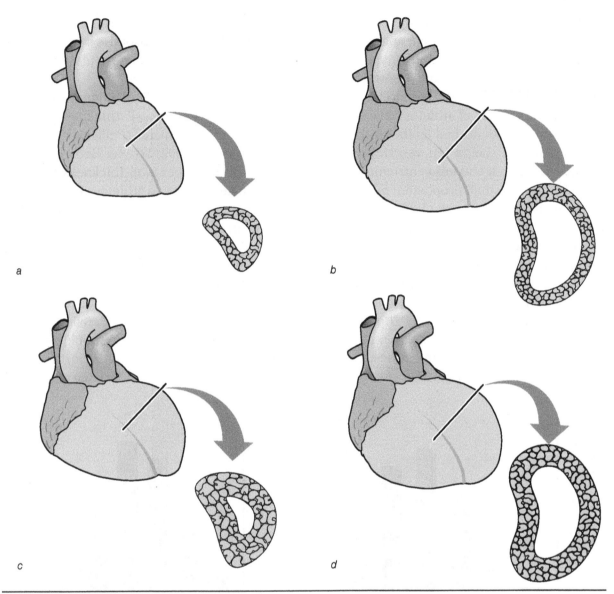

Figure 6.2 The differential cardiac adaptation associated with training stimulus. *(a)* Heart of a normal individual. *(b)* Heart of an endurance trained athlete. *(c)* Heart of a strength trained athlete. *(d)* Heart of a combined endurance and strength trained athlete.

Whyte, George, Sharma et al., 2004; see figure 6.2). A recent review has suggested limitations in the Morganroth hypothesis (Naylor et al., 2008). There is a paucity of research data to support the afterload-induced cardiac adaptation in strength-trained athletes, and endurance-trained athletes appear to display significant cardiac remodeling, whereas their strength-trained counterparts demonstrate limited cardiac adaptation.

Cardiac dimensions in athletes are slightly larger than those of nonathletic matched controls, and a large overlap is often observed. Although the difference is small, it does reach statistical significance. Athletes exhibit a 15 to 20% larger left ventricular wall thickness and 10% larger left ventricular cavity size compared with nonathletes. These modest increases result in a marked increase in left ventricular mass in the region of 50% (Maron, 2002). Left atrial enlargement is also observed alongside ventricular enlargement in athletes (Basavarajaiah et al., 2006). Ethnic differences exist in the cardiac remodelling to physical training: athletes of Afro-Caribbean origin exhibit greater wall thickness (the upper limit is 16 mm) and markedly different ECGs compared with Caucasian athletes (Basavarajaiah et al., 2008). Despite significant increases in left ventricular mass, normal or enhanced indices of left ventricular systolic and diastolic function are observed in AH (Whyte, George, Neville et al., 2004; Whyte, George, Sharma et al., 2004).

The vast majority of athletes have cardiac dimensions within normal limits for the general population (i.e., a left ventricular wall thickness <12 mm and left ventricular cavity size <55 mm). A small proportion of athletes, however, have a left ventricular wall thickness and more commonly a left ventricular cavity size exceeding predicted normal limits. In this group of athletes, cardiac dimensions may be similar to those seen in patients with morphologically mild hypertrophic and dilated cardiomyopathy, respectively (Pelliccia et al., 1991; Sharma et al., 2002; Whyte, George, Sharma et al., 2004). The differentiation between physiological cardiac enlargement

(AH) and cardiomyopathy is crucial considering that the cardiomyopathies are the most common cause of exercise-related sudden death (Sharma et al., 1997).

Studies have identified upper normal limits for left ventricular wall thickness in adult and adolescent athletes (Pelliccia et al., 1991; Sharma et al., 2002; Whyte, George, Sharma et al., 2004). Upper wall thickness limits for adult male and female athletes are 14 mm and 12 mm, respectively. In adolescent athletes a left ventricular wall thickness >12 mm warrants further investigation. Left ventricular cavity size in athletes more commonly exceeds normal limits; upper limits for male and female athletes are 66 mm and 60 mm, respectively (Whyte, George, Sharma et al., 2004; see figure 6.3). Values in excess of these should be viewed with caution and should prompt further investigation to identify the underlying cause.

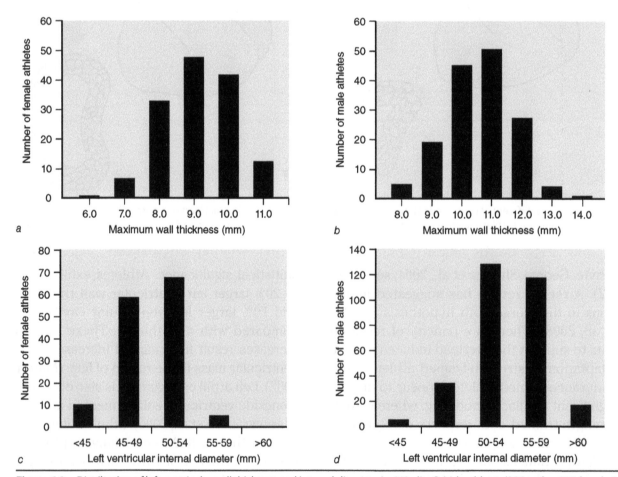

Figure 6.3 Distribution of left ventricular wall thickness and internal diameter in 442 elite British athletes (306 males, 136 females).

With kind permission from Springer Science+Business Media: *European Journal of Applied Physiology,* The upper limit of physiological cardiac hypertrophy in elite male and female athletes: The British experience, Vol. 92, 2004, pgs. 592-597, G. Whyte, K. George, S. Sharma, S. Firrozi, N. Stephens, R. Senior, and W. McKenna. © Springer Science.

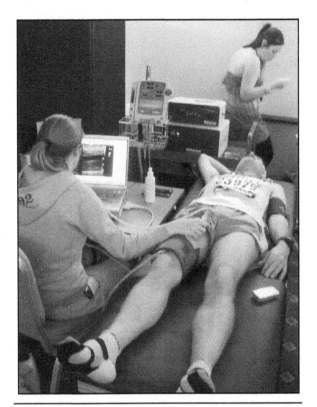

Figure 6.4 The evaluation of vascular structure and function in an athlete. Vascular M-mode and Doppler ultrasound are being used to measure femoral artery structure and function at rest and in response to an ischemic challenge (flow-mediated dilatation, FMD).

In addition to central cardiac changes, athletes present with significant changes in vascular structure and function. An increased arterial diameter and enhanced vascular endothelial function are observed in the conduit arteries of athletes (i.e., brachial, femoral, popliteal) (Green et al., 2004; see figure 6.4).

ELECTROCARDIOGRAM (ECG) OF AN ATHLETE

Common ECG manifestations of the athlete's heart include sinus bradycardia, sinus arrhythmia, voltage criteria for left ventricular enlargement, and early repolarization changes such as tall T waves and concave ST-segment elevation. Incomplete right bundle branch block (rSr' pattern in V_1, possibly reflecting right ventricular enlargement) is also relatively common. First-degree heart block (PR interval >0.2 s) and Mobitz type I second-degree AV block (see chapter 4) are also recognized findings, although higher degrees of AV block are rare. Minor T-wave inversions are observed and usually confined to the right chest leads (V_1, V_2, V_3), but ST-segment depression or deep (>0.3 mV) T waves are uncommon. Pathological Q waves and left bundle branch block are not features of the athlete's heart (see table 6.1).

Figure 6.5 on page 120 shows an ECG from an elite endurance athlete demonstrating resting bradycardia, first-degree heart block, incomplete right bundle branch block, and left ventricular hypertrophy. Significant ethnic differences exist in the resting ECG of athletes (Basavarajaiah et al., 2008). Figure 6.6 on page 120 shows a normal resting ECG from an Afro-Caribbean athlete demonstrating significant LVH (Sokolow criteria = 52 mm), dome-shaped ST-segment elevation, and T-wave inversion

Table 6.1 Common and Uncommon Findings on Resting 12-Lead ECG in Athletes

Common	Uncommon
Sinus bradycardia/arrhythmia	
First-degree AV block	Second (Mobitz type II and advanced) and third degrees AV block
QT and QRS elongation	QT >0.44 ms
LVH (Sokolow-Lyon)	LVH (Romhilt-Estes) in females
LA/RA enlargement	
ST-segment elevation	ST depression in any lead
Tall T waves	Deep T-wave inversion
	Minor T-wave inversion (<16 yrs)
Partial RBBB	

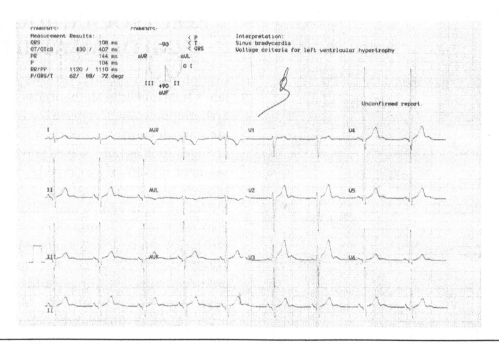

Figure 6.5 An ECG from an elite Caucasian endurance athlete showing resting bradycardia, first-degree heart block, incomplete right bundle branch block, and left ventricular hypertrophy.

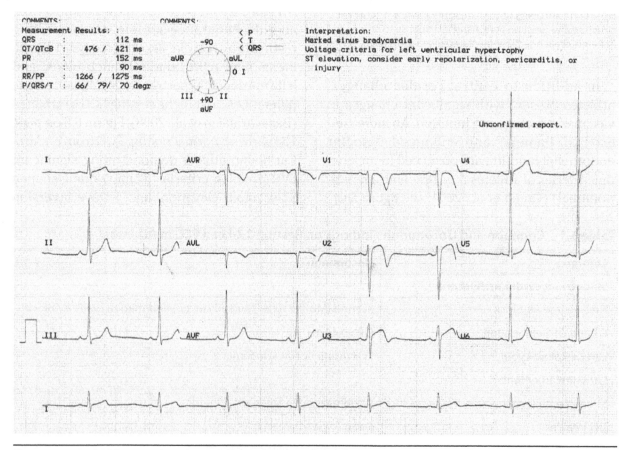

Figure 6.6 An ECG from an elite Afro-Caribbean athlete demonstrating significant LVH (Sokolow criteria = 52 mm), dome-shaped ST-segment elevation, and T-wave inversion in the right chest leads (V_1, V_2, V_3).

in the right chest leads (V_1, V_2, V_3). Deep T-wave inversions may also be present in Afro-Caribbean athletes but are rarely observed in Caucasian athletes.

During exercise some of these anomalies disappear. Heart block and T-wave inversions are often normalized during exercise.

ARRHYTHMIAS AND THE ATHLETE

Cardiac arrhythmias in athletes range from the benign and asymptomatic to the symptomatic and potentially life threatening. Supraventricular and ventricular extrasystoles are common and are usually of no clinical significance. The high vagal tone associated with physical training may result in athletes being more susceptible to certain bradyarrhythmias. Asymptomatic bradyarrhythmias such as sinus bradycardia, nodal bradycardia, and Mobitz second-degree AV block (Mobitz type I) are common among athletes and are due to the high vagal tone associated with intense physical training. Higher degrees of AV block and supraventricular tachyarrhythmias are uncommon in athletes. In a small number of athletes, high vagal tone may predispose to atrial fibrillation (Whyte, Stephens, Budgett et al., 2004b).

Potentially life-threatening ventricular arrhythmias are uncommon in athletes and are generally associated with underlying structural heart disease, coronary artery disease, or ion channnelopathies. In these circumstances participation in sport of high and moderate intensities is contraindicated. In a few cases, ventricular tachycardia may occur in the absence of these predisposing substrates and be amenable to treatment with electrophysiological radiofrequency ablation (e.g. right ventricular outflow tract ventricular tachycardia, RVOT-VT; Whyte, Stephens et al., 2008). Athletes presenting with complex supraventricular and ventricular arrhythmias should be referred for a full cardiovascular workup to identify the underlying cause.

SYNCOPE AND THE ATHLETE

Unexplained syncope in an athlete is a potentially ominous symptom that requires a thorough cardiovascular evaluation. The most frequent cause is vasovagal syncope, commonly associated with neurally mediated mechanisms, but the condition may be compounded by dehydration and hyperthermia. Presyncope and orthostatic intolerance are common following exhaustive exercise and in a small number of predisposed people, can lead to syncope. The mechanisms underlying these postexercise responses are associated with cardiac factors (heart rate and stroke volume), vascular factors (total peripheral resistance, TPR), or both (Privett et al., 2008; Whyte, Stephens, Budgett et al., 2004a; see figure 6.7).

Although the vast majority of cases of syncope occur in the absence of underlying cardiac pathology, structural heart disease and other causes should be eliminated before considering vasovagal syncope as the etiology (see table 6.2). Integrated cardiopulmonary stress testing is useful and should be performed while recording the ECG during the

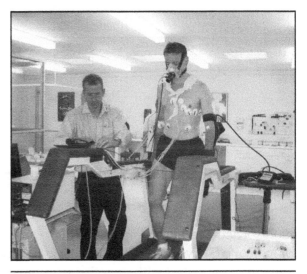

Figure 6.7 Evaluating an athlete presenting with vasovagal syncope using integrated cardiopulmonary stress testing to evaluate the cardiovascular response pre-, during, and postexercise.

Table 6.2   Causes of Syncope

Vasovagal	Emotional faint
	Carotid sinus syncope
	Coughing
	Swallowing, defecation, urination
	Airway stimulation
	Postexercise
Orthostatic and vascular	Idiopathic orthostatic hypotension
	Shy-Drager syndrome
	Diabetic neuropathy with orthostatic hypotension
	Drug-induced orthostasis
Cardiac arrhythmia	Sinus node dysfunction
	Atrioventricular conduction system disease
	Paroxysmal supraventricular tachycardia
	Paroxysmal ventricular tachycardia
	Cardiac implant malfunction
Structural cardiac disease	Valvular disease
	Myocardial infarction
	Obstructive cardiomyopathy
	Subclavian steal syndrome
	Pericardial disease
	Pulmonary embolus
	Primary pulmonary hypotension
Neurological	Cerebrovascular (e.g., vertebrobasilar disease)
	Central nervous system substrate disorder (e.g., seizure disorder, subarachnoid hemorrhage)
Noncardiovascular	Hypoglycemia
	Volume depletion
	Hypoxemia
	Hyperventilation
	Panic attack
	Hysteria

Adapted from D. Wang, S. Sakaguchi, and M. Babcoack, 1997, "Exercise induced vasovagal syncope," *Physician SportsMedicine* 25: 64-74.

mode of activity that precipitates presyncope or syncope (Whyte et al., 1999). Athletes with syncope associated with exercise who have a fully negative cardiac evaluation can safely participate in sports of all types and intensities. The exception to this rule are those with recurrent syncope who participate in sports in which even a transient loss of consciousness may be hazardous (e.g., high-speed or water sports).

Traditional pharmacotherapy for vasovagal syndrome includes β_1-andrenergic blocking agents, antiarrhythmics, and plasma volume expanders. These groups of drugs are currently prohibited by international governing bodies of sport. In addition, care must be taken when prescribing drugs that have a negative inotropic action. Strategies to reduce or eliminate vasovagal symptoms are targeted at maintaining blood pressure and venous return postexercise and include cool-downs, coughing, muscle tensing, and leg crossing (Krediet et al., 2002).

SUDDEN CARDIAC DEATH (SCD) IN ATHLETES

Sudden cardiac death in a young athlete is defined as a nontraumatic, nonviolent, unexpected death due to a cardiac cause. Sudden death can occur during exercise or at rest. The time course of death remains equivocal ranging from during or immediately postexercise to 24 hours postexercise. Despite a lack of consensus agreement, previous studies have reported a tenfold increase in the incidence of cardiovascular events in those with cardiovascular disease during exercise (Seto, 2003). The relative risk of exercise leading to sudden cardiac death in young athletes has been estimated at 2.5 times higher for an adverse event (Corrado et al., 1998). The sudden death of a young athlete is an uncommon event with an estimated prevalence between 1 in 100,000 and 1 in 300,000; death rates are nine times higher in males than females (Papadakis et al., 2008). Data from the National Federation of State High School Associations (USA) estimate that 10 to 25 deaths in young athletes (<30 years) occur per year (Van Camp et al., 1995). The exact prevalence of sudden cardiac death remains unclear because of the paucity of well-controlled studies. Most studies to date have relied on self-reporting of physicians and media accounts of deaths.

Despite the relatively low prevalence, the sudden death of an athlete has a profound effect not only on the athlete's immediate family but also on teammates, the local community, and the sport as whole. Indeed, the sudden cardiac death of a young athlete is often highly publicized leading to ramifications nationally and internationally.

Sudden cardiac death in young (<35 years) athletes is associated with a small number of inherited, congenital, and acquired cardiac diseases (see table 6.3). Data from the UK, Italy, and the USA suggest that cardiomyopathies are the most common

Table 6.3   Causes of Exercise-Related Sudden Cardiac Death in Young Athletes

Common	Uncommon
Cardiomyopathies including: 　Idiopathic left ventricular hypertrophy 　Hypertrophic cardiomypathy (HCM)*	Myocarditis
Arrhythmogenic right ventricular cardiomypathy (ARVC)**	Coronary artery disease (CAD)
Ion channelopathies including: 　Long QT syndrome 　Brugada syndrome	Marfan's syndrome Mitral valve prolapse (MVP) Wolff-Parkinson-White syndrome (WPW) Aortic stenosis

*Most common cause of Exercise Related Sudden Cardiac Death (ERSCD).

**Most common cause of ERSCD in the Veneto region of northern Italy.

Adapted from S. Sharma, G. Whyte, and W.J. McKenna, 1997, "Sudden cardiac death in young athletes - fact or fiction?" *British Journal of Sports Medicine* 31(4): 269-276.

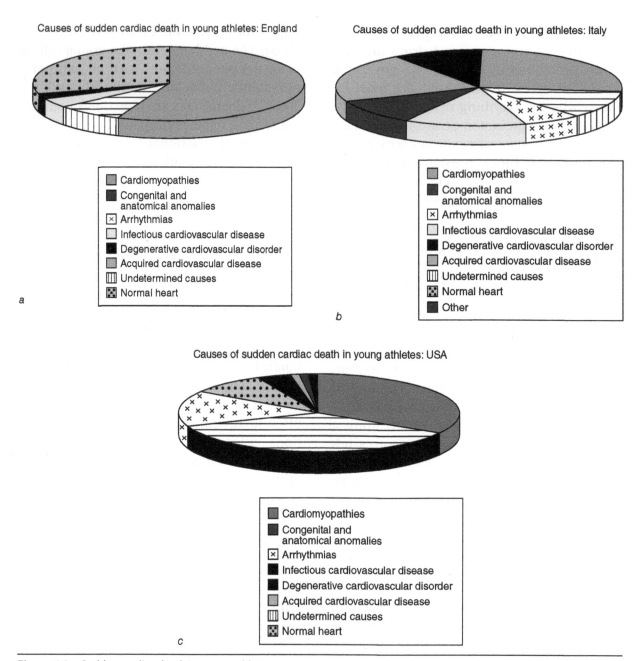

Figure 6.8 Sudden cardiac death in young athletes.

Data from England and USA: Adapted from B. Maron et al. 2007, "Recommendations and considerations related to pre-participation screening for cardiovascular abnormalities in competitive athletes: 2007 update," *Circulation* 115: 1643-1655: Italy: Adapted from D. Corrado et al., 2006, "Trends in sudden cardiovascular death in young competitive athletes after implantation of a preparticipation screening programme," *JAMA* 296: 1593-1601.

cause of SCD in young athletes (see figure 6.8). Hypertrophic cardiomyopathy accounts for around 50% of sudden cardiac deaths in the USA (Maron et al., 1996), whereas arrhythmogenic right ventricular cardiomyopathy (ARVC) is the most common cause in the Vento region of northern Italy. The low prevalence of HCM deaths in the Italian data, however, is likely associated with the success of preparticipation screening resulting in a bias toward non-HCM causes of SCD. Other common causes of sudden cardiac death in young athletes include anomalous coronary arteries and the ion channelopathies including Brugada and long QT syndromes (Papadakis et al., 2008).

The incidence of sudden death in older athletes (>35 years) has been reported as 1 death per 396,000 hours of jogging (Thompson et al., 1982), but appears to be lower for men with higher levels of habitual physical activity (ACSM, 1995). No evidence currently exists for death rates in women. Coronary artery disease is the most common condition leading to exercise-related sudden cardiac death. The increased metabolic and physiologic demands experienced during exercise lead to an increased risk of a cardiac event. Sudden death is often observed in highly conditioned older athletes as a result of occult coronary atheroma. Those older athletes at higher risk often have more than one recognized risk factor including smoking, a family history of myocardial infarction under the age of 55, hypertension, and hypercholesterolemia.

PREPARTICIPATION SCREENING

The physiological adaptation of the heart to physical training may mimic pathology (Sharma et al., 1999; Whyte, George, Sharma et al., 2004). Differentiating physiological from pathological changes in cardiac structure and function is crucial to reducing the prevalence of SCD in young athletes. The causes of SCD together with the increased incidence of events during exercise indicate that adequate cardiovascular screening and evaluation are important to identify those with underlying cardiovascular disease prior to the commencement of rigorous training and competition. A full understanding of athletes' ECG at rest and during exercise facilitates the diagnosis of pathologies, helps identify at-risk athletes, and is the cornerstone of a successful pre-participation screening program.

In 1996 the American Heart Association (AHA) developed consensus recommendations for the cardiovascular screening of student-athletes as part of a comprehensive preparticipation physical examination. This was incorporated into the preparticipation physical evaluation guidelines in the USA (Smith et al., 1997). This system of preparticipation screening does not exist for sport in the UK. Indeed, because of the large differences in design and content of preparticipation screening among sports in the UK (Batt, et al., 2004), a national standard for preparticipation medical evaluation, including cardiovascular screening, would enhance the quality of care offered to athletes. Although limited evidence from the Italians suggests that preparticipation screening reduces the incidence of sudden cardiac death (Thiene et al., 1999), the efficacy of such programs remains to be fully elucidated because of the absence of systematic cardiovascular screening programs in other countries, including the UK.

In general, a comprehensive cardiovascular screen should include family history, medication and drug use history, blood pressure, pulse, recognition of Marfan's syndrome, auscultation of the heart (unnecessary when echocardiography is employed), ECG, and echocardiography. A 12-lead integrated cardiopulmonary stress test may be a valuable tool in a differential diagnosis and is often part of an athlete's support program (Whyte et al., 1999).

Often, the preparticipation screening debate surrounds the issues of cost and time. Because echocardiography and cardiopulmonary stress testing add significantly to the time and cost of screening, these tools are often employed as follow-up tests after the initial screening, along with more specialized tests including ambulatory ECG monitoring, implantable loop recorders, and cardiac magnetic resonance imaging.

In addition to debating the efficacy of preparticipation screening in reducing the incidence of sudden cardiac death, experts also debate the value of including a resting 12-lead ECG. Although the Europeans have adopted a preparticipation screening approach that includes a resting 12-lead ECG, the USA remains opposed to its inclusion based on a perceived absence of evidence supporting an improved diagnostic capacity above that offered by family and symptom history questionnaires and physical examinations. Recent evidence, however, reinforces the importance of including a resting 12-lead ECG in the initial preparticipation examination because of the poor specificity and sensitivity of questionnaires alone (Wilson et al., 2008).

For older athletes, education programs should be aimed at increasing awareness of warning symptoms, including chest pain, palpitation, and syncope (fainting). Those over the age of 40 should undertake severe exercise cautiously, particularly if they have risk factors. Older people intending to train for the first time or following a prolonged period of inactivity should seek medical advice before doing so. Those presenting with symptoms suggestive of coronary artery disease during exercise—including chest pain, unusual shortness of breath, dizziness, and palpitations—should be fully evaluated prior to recommencing training.

A cause of SCD in people free of cardiac disease is blunt chest impact produced by either a projectile or a collision with another athlete (commotio cordis). Three determinants of a commotio cordis have been identified:

- A relatively low-energy chest impact directly over the heart
- Precise timing of the blow to a narrow 15 ms segment of the cardiac cycle vulnerable to potentially lethal ventricular arrhythmias, just prior to the T-wave peak
- A narrow, compliant chest wall, typical of young children

The majority of commotio cordis events are fatal; however, a small number of athletes survive, supporting the importance of early recognition and prompt institution of CPR and defibrillation (Maron et al., 2002).

KEY POINTS

▶ Physical training results in adaptations of the structure and function of the heart that are collectively known as "The Athlete's Heart" (AH). Commonly observed changes include: resting bradycardia (< 60 beats·min⁻¹); left ventricular enlargement (increased wall thickness and chamber dimensions); a variety of resting ECG changes and; additional heart sounds on auscultation.

▶ The type and magnitude of changes observed vary by gender (male vs. female) and race (i.e. Blacks).

▶ The changes observed in AH can mimic those observed in diseased states associated with sudden cardiac death in young athletes including: heart muscle disorders (HCM, ARVC); electrical conduction abnormalities (LQTS, Brugada's) and; structural abnormalities (coronary artery anomalies).

▶ Sudden cardiac death (SCD) in athletic populations is an infrequent but regular occurrence. Pre-participation screening may be valuable in the early identification of athletes' at risk, however, more data is required to support the efficacy of screening programs.

Athlete Case Studies

The ECG at rest and during exercise is used in a number of clinical and performance settings to assist in the diagnosis of diseased states. This chapter presents six case studies that demonstrate the variety and versatility of the 12-lead ECG in clinical settings. Each case study presents findings from a single athlete or multiple athletes with the same problem, discusses relevant literature, and provides conclusions in the form of a take-home message. The clinical problems addressed include spontaneous atrial fibrillation, exercise-induced vasodepressor syncope, long QT syndrome, right ventricular outflow tract ventricular tachycardia (RVOT-VT) versus arrhythmogenic right ventricular cardiomyopathy (ARVC), and the impact of ultra-endurance exercise on cardiac structure and function.

CASE ❶ STUDY

Spontaneous Atrial Fibrillation in a Freestyle Skier

BACKGROUND

In October 2002, a 19-year-old white male freestyle skier presented to the British Olympic Medical Centre for a routine physiological assessment. Following the completion of a standard preexercise questionnaire and the signing of an informed consent form, it was established that the athlete had no past history of note; including no family history of heart disease, was a nonsmoker, drank alcohol occasionally, and was not on drug therapy at the time of investigation.

Routine physiological assessment included an incremental treadmill test to volitional exhaustion to ascertain $\dot{V}O_2$max, during which expired gases were collected at the mouth (Oxycon Alpha, Viasys, UK) and heart rate was monitored using telemetry (Polar Electro Oy, Finland). Following the test, the athlete's heart rate failed to fall as expected. Further, the athlete complained of chest pain and shortness of breath. The athlete was consequently attached to a 12-lead ECG that revealed atrial fibrillation with a ventricular rate of 155 beats·min[-1] (see figure 7.1). The athlete was monitored for two hours, during which time the AF persisted. However, the athlete became mildly compromised indicated by a fall in blood pressure and a worsening discomfort associated with chest pain and shortness of breath. Following an echocardiographic interrogation to determine the absence of structural heart disease and a normal cardiac function, the athlete was

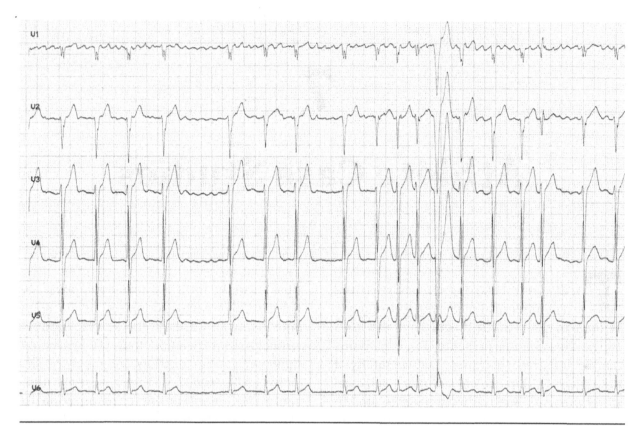

Figure 7.1   Postexercise ECG demonstrating atrial fibrillation with a ventricular rate of 155 beats·min⁻¹.

admitted to the Coronary Care Unit (Resuscitation Council [UK], 2004). The AF was successfully cardioverted with flecanaide (150 mg over 30 minutes), and following 24 hours of observation, the athlete was discharged.

On follow-up examination four weeks later, the athlete was in sinus rhythm and had normal venous pulse and normal heart sounds. Resting blood pressure was 124/80 mmHg. Echocardiography demonstrated normal intracardiac dimensions with normal systolic function of both left and right ventricles, and 24-hour ECG demonstrated sinus rhythm throughout. During integrated cardiopulmonary exercise stress testing, the athlete completed 1 minute, 20 seconds of level 5 of a Bruce protocol stopping because he reached maximal exercise capacity. The athlete achieved a V̇O$_2$max of 53.5 ml/kg/min (150% of age and gender predicted) and a maximum heart rate of 201 beats·min⁻¹ (100% of age predicted). Blood pressure and heart

rate response during exercise were normal. There were no inducible ECG changes or arrhythmias during or postexercise.

During consultation the athlete admitted that he had consumed 12 units of alcohol (an excessive amount compared with his normal consumption) two days prior to the initial physiological assessment. The observed AF was diagnosed as a lone episode of alcohol-induced AF, and the athlete was counseled regarding alcohol consumption and vigorous exercise. No further action was taken, and the athlete has reported no further occurrence of symptoms indicative of AF.

DISCUSSION

In the absence of a preexercise ECG and ECG monitoring throughout the exercise test, it is difficult to ascertain whether the observed AF was present prior to exercise. The athlete, however, was symptomless with a heart rate of around 70 beats·min⁻¹ at the onset of exercise. It is, therefore, likely that the observed AF represented the first presentation of the arrhythmia. AF may be related to a number of acute temporary causes including drug use (anabolic steroids; Sullivan et al., 1999), surgery, electrocution, MI, pericarditis, myocarditis, pulmonary embolism or other pulmonary diseases, hyperthyroidism, or other metabolic disorders (ACC/AHA/ESC, 2001). Most pertinent to this case study, however, is the effect of alcohol on the presentation of AF. Alcohol may increase the irritability of the atria (a situation often called "holiday heart syndrome") resulting in the potential for catecholamine-induced AF associated with exhaustive exercise.

Supraventricular arrhythmias are uncommon in athletes, with the exception of AF, which may occur more frequently than in the general population as a result of a high vagal tone (Furlanello et al., 1998; Link et al., 2001). Vagally mediated AF appears to be more prevalent in male athletes (Furlanello et al.,

1998) and is frequently associated with "lone AF" (ACC/AHA/ESC, 2001). The term *lone AF* is used to describe an isolated occurrence and generally applies to young people without clinical or echocardiographic evidence of cardiovascular disease. In the absence of structural cardiac disease, the likely etiology of AF is atrial tachycardia originating at the insertion of the pulmonary veins into the left atrium, which degenerates into AF (Haissaguerre et al., 1998).

The ventricular response to AF depends on the electrophysiological properties of the AV node and the level of vagal and sympathetic tone. Extremely rapid ventricular rates (>200 beats·min^{-1}) suggest the presence of an accessory pathway. The athlete in this case study did not have an accessory pathway as evidenced by a ventricular rate of 155 beats·min^{-1}, together with the normal PR interval and absence of delta waves on the resting ECG postcardioversion (see figure 7.2). Analysis of the resting ECG demonstrates a QS

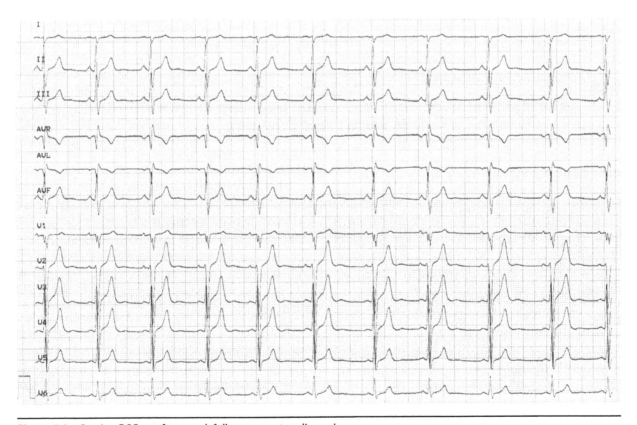

Figure 7.2　Resting ECG at a four-week follow-up postcardioversion.

pattern in V_1, indicative of incomplete right bundle branch block, together with repolarization abnormalities, including ST-segment elevation and peaked T waves. These findings are common in athletes and represent a physiological adaptation to training (Sharma et al., 1999).

Because the incidence of thromboembolism is very low in athletes presenting with self-limiting episodes of AF (Link et al., 2001), antiarrhythmic drugs to prevent recurrence are usually unnecessary, unless AF is associated with severe symptoms, structural heart disease, or both (ACC/AHA/ESC, 2001). For those athletes with symptoms, beta-blockers or calcium channel blockers may provide some relief; however, beta-blockers are currently on the banned drug list for many sports. The use of RF ablation of the atrial tachycardia focus may be warranted for athletes with recurrent or troublesome episodes of paroxysmal AF.

Cardiac arrhythmias in athletes range from the benign and asymptomatic to the symptomatic and potentially life threatening (Link

et al., 2001). The high vagal tone associated with physical training may result in athletes being more susceptible to certain arrhythmias including AF. It is important to determine which athletes with arrhythmia require follow-up and therapy, and which can safely return to training and competition.

TAKE-HOME MESSAGE

Small volumes of alcohol can have a relatively large effect on an elite athlete. Athletes should be counseled about the potential dangers of consuming unusually large quantities of alcohol prior to exhaustive exercise. ECG monitoring during physiological assessment is crucial to improve the care and safety of the athlete, and should be adopted as standard. Although AF is more prevalent in athletes, it remains uncommon. Care must be taken in investigating the ancestry of an athlete presenting with AF to eliminate the possibility of familial AF.

Adapted from *British Journal of Sports Medicine,* "Spontaneous atrial fibrillation in a freestyle skier," G. Whyte, N. Stephens, R. Budgett, S. Sharma, R. Shave, and W. McKenna, 38(2): 230-232, 2004.

CASE **2** STUDY

Exercise-Induced Vasodepressor Syncope in an Elite Rower: A Treatment Dilemma

BACKGROUND

In March 2002, a 25-year-old white female presented to the British Olympic Medical Centre following bouts of syncope and presyncope. The athlete was an international rower competing in the single sculls event, and was training an average of 10 sessions per week. Training sessions included rowing and weight training across the full range of intensities. The presenting symptoms included two episodes of syncope following 2,000 m maximal tests on a rowing ergometer. Separately, the patient noted a tendency for presyncope and syncope during medical procedures. There was no past

history of note including no family history of heart disease. The athlete was a nonsmoker, drank alcohol occasionally, and was not on drug therapy at the time of investigation.

On examination, the athlete was in sinus rhythm with marked bradycardia (~around 40 beats·min^{-1}) and sinus arrhythmia. All other measures, including PR and QT interval, were within normal limits. Resting blood pressure was 110/70 mmHg. Echocardiography demonstrated normal intracardiac dimensions with normal systolic function of both the left and right ventricles. Mild pulmonary valve regurgitation was noted (inaudible clinically).

During integrated cardiopulmonary exercise stress testing, the athlete completed level 5 of a Bruce protocol stopping at maximal exercise capacity. The athlete achieved a $\dot{V}O_2$max of 60.6 ml/kg/min (185% of age and gender predicted) and a maximum heart rate of 186 beats·min^{-1} (95% of age predicted). Blood pressure and heart rate response during exercise were normal. There were no inducible ECG changes or arrhythmias. On cessation of exercise, however, there was a precipitate fall in heart rate (186 to 100 beats·min^{-1}) and blood pressure (200/90 to 110/60 mmHg).

The athlete was diagnosed with mild exercise-induced vasovagal syndrome, a condition considered benign. However, because the rower was an international-caliber single-scull rower, the potential for a syncopal episode on water and subsequent immersion and drowning raised a treatment dilemma.

DISCUSSION

Syncope is the transient loss of consciousness and postural tone. It rarely occurs during exercise, but when it does, it is often an ominous sign of structural heart disease (Sneddon et al., 1994). In contrast, however, recurrent idiopathic syncope has been described following exercise in patients with normal hearts (Grubb et al., 1993) and has been reported to be particularly prevalent in endurance-trained athletes (Levine et al., 1991).

The pathophysiological mechanisms of exercise-induced vasodepressor syncope are incompletely understood. Early work examining baroreceptor control in athletes suggested that blood pressure control was impaired as a result of a depressed carotid baroreflex sensitivity (Stegmann et al., 1974). More recent work (Levine et al., 1991) has identified stroke volume as being of equal importance in the "triple product" of blood pressure control (HR $\times$ SV $\times$ TPR). Levine and colleagues (1991) suggested that endurance athletes have more compliant, distensible ventricles associated with a chronic volume load during training

and, therefore, a steeper slope of the Frank-Starling curve relating left ventricular filling pressure to stroke volume than nonathletes. The increase in chamber compliance, steep Frank-Starling curve, and abnormal blood pressure response, facilitating the delivery of a large volume of blood to the exercising muscle, may be beneficial to the performing athlete. These conditions may, however, be a disadvantage during orthostasis, resulting in a large decrease in stroke volume when filling pressure is reduced (Grubb et al., 1993).

The athlete in this case study demonstrated normal echocardiographic findings, no inducible arrhythmia during exercise, and a normal resting ECG, suggesting the absence of a pathological origin of the reported syncopal episodes. Recent evidence (Colivicchi et al., 2002) suggests that exercise-related syncope is not associated with adverse outcomes in competitive athletes with fully negative cardiovascular workups. Therefore, in the absence of cardiac disease, the athlete should not be barred from competition. Indeed, the athlete in this case study refused to cease training and competition.

Traditional pharmacotherapy for vasovagal syndrome includes β_1-andrenergic blocking agents, antiarrhythmics, and plasma volume expanders (Grubb et al., 1993). A number of factors must be taken into consideration when prescribing such drugs for the rower in this case study and elite athletes in general. The international governing body for rowing (FISA) has banned the use of plasma volume expanders (both in and out of competition) and beta-blockers (in competition). In addition, care must be taken when prescribing drugs that have negative inotropic actions. Submaximal and maximal exercise capacity are reduced in normotensive, asymptomatic people following beta-blocker administration (Van Baak, 1988). Thus, the practitioners were unable and unwilling to control the exercise-induced vasodepressor syncope in the rower in this case study through pharmacotherapy.

In the absence of traditional drug therapy, and given the understandable unwillingness of the rower to abstain from training and competition, a cool-down program specifically to reduce the potential for the development of vasovagal symptoms was designed. The program was targeted at maintaining blood pressure and venous return by having the athlete continue to exercise; rather than stop abruptly following exercise, particularly high-intensity training and competition. Since adopting the new program, the rower has reported only moderate symptoms of presyncope and no syncopal episodes following exercise including a 2,000 m time trial.

TAKE-HOME MESSAGE

Exercise-induced vasodepressor syncope in elite athletes is not uncommon and is more prevalent in endurance-trained athletes. In the presence of a negative cardiovascular workup, exercise-induced vasodepressor syncope is considered benign. Pharmacotherapy may not be a treatment option, particularly in elite, competitive athletes. Although this condition may be seen as benign, there is a potential for syncope-related traumatic injuries and, in the case of athletes on or in water, death. Comprehensive training strategies to reduce the potential for syncope should be given to the athlete including continuation of exercise and lower body positive pressure maneuvers postexercise. In addition, teammates and support staff should be aware of the situation.

Adapted from *British Journal Sports Medicine*, "Exercise induced vasodepressor syncope in an elite rower: A treatment dilemma," G. Whyte, N. Stephens, R. Budgett, S. Sharma, R. Shave, and W. McKenna, 38(1): 84-85. © 2004 with permission from BMJ Publishing Group Ltd.

CASE **3** STUDY

Prevalence and Significance of an Isolated Long QT Interval in Elite Athletes

BACKGROUND

Congenital long QT syndromes (LQTS) are a recognized cause of adrenergic-mediated polymorphic ventricular tachycardia (VT) and have been implicated in exercise-related sudden cardiac deaths in young athletes (Basavarajaiah et al., 2007). The identification of a prolonged QT interval corrected for heart rate (QT_c) in an athlete is recommended as a reason for disqualification from competitive sport involving strenuous exertion (Pelliccia et al., 2006; Sharma et al., 1999) even though an isolated prolonged QT_c interval *per se* does not fulfil the definition of congenital long QT syndrome.

It is well documented that athletes exhibit a more pronged QT_c interval compared with nonathletes (Pelliccia et al., 2000) raising the possibility of an erroneous false positive diagnosis of LQTS. Although numerous electrocardiographic studies relating to athletes exist (Biffi et al., 2002; Bjornstad et al., 2006; Pelliccia et al., 2000), the prevalence and, more important, the significance of an isolated, prolonged QT_c interval in athletes has never been evaluated.

The aim of this study was to identify the prevalence of a prolonged QT_c interval in a large cohort of elite athletes and evaluate its significance using Holter monitoring, exercise testing, cardiovascular evaluation of first-degree relatives, and genetic testing in consenting athletes.

Setting

The death of several professional athletes from inherited structural or electrical disorders has resulted in many professionals calling for the screening of all athletes for potentially sinister cardiovascular disorders prior to selection for competition (Monnig et al., 2006). However, in the UK, as in many other developed countries, a comprehensive screening program is not possible because of constraints on financial resources, personnel, and administrative infrastructure. Nevertheless, certain sporting bodies such as the British Lawn Tennis Association (LTA), the premiership football and rugby leagues, and the national swimming and boxing squads have adopted self-financed mandatory preparticipation screening programs that include histories, physical examinations, 12-lead ECGs, and echocardiography. Further investigations are undertaken if necessary.

Athletes

Between 1996 and 2006, 2,000 elite athletes between the ages of 14 and 35 (mean 20.24 years) were evaluated. Of those, 1,400 (70%) were male athletes and 600 (30%) were female athletes. All athletes had structurally and functionally normal hearts as seen on two-dimensional echocardiography. The athletes participated in 15 different sports, but the vast majority of the study group (71%) were football players, rugby players, tennis players, or swimmers (see table 7.1). All athletes competed at least at a regional level, and approximately 50% of them were playing at a national level during the study period. Written consent was obtained from those over age 16, and from a parent or guardian for those under age 16.

Health Questionnaire

All athletes filled out a health questionnaire that addressed cardiovascular symptoms, with particular emphasis on the relationship with physical exertion; past medical history; regular medications; and a family history of inherited cardiovascular conditions, premature SCD, epilepsy, or unexplained deaths in first-degree relatives from drowning or road traffic accidents.

Table 7.1 Number of Athletes in Various Sports

Sport	Number of athletes	Percentage (%)
Football	520	26
Tennis	450	22.5
Rugby	256	12.8
Swimming	202	10.1
Rowing	88	4.4
Cycling	64	3.2
Athletics	64	3.2
Badminton	54	2.7
Netball	52	2.6
Basketball	52	2.6
Triathlon	52	2.6
Boxing	42	2.1
Hurling	40	2
Fencing	32	1.6
Speed skating	32	1.6
Total	**2,000**	

Electrocardiography

Standard 12-lead ECGs were recorded using a Marquette Hellige recorder (Milwaukee, Wisconsin) with a paper speed of 50 mm/s and an amplification of 0.1 mV/mm. The QT interval was measured manually in lead II from the onset of the QRS complex to the end of the T wave, defined as the intersection of the isoelectric line and the tangent of the maximal downward limb of the T wave (Moss, 1993). The U wave was excluded in the measurement of the QT interval, except when the T wave was biphasic or in the presence of T-U complexes, when identification of the termination of the T wave was difficult. In such cases the U wave was included if it exceeded 50% of the T-wave amplitude (Funck-Brentano and Jaillon, 1993). The QT interval was measured for three to five consecutive beats and was averaged. The QT interval was corrected for the heart rate (QT_c) using Bazett's formula (Bruce, 1971).

FURTHER INVESTIGATIONS

Athletes with a QT_c >440 ms were detrained for six weeks and underwent a repeat ECG and further assessment with 48-hour ECG monitoring and exercise stress testing to identify additional phenotypic features of congenital LQTS.

48-Hour ECG

Athletes were encouraged to continue with usual (nonathletic) life activities while undergoing ECG monitoring. The 48-hour ECG monitors were analyzed specifically for episodes of polymorphic VT.

Exercise Stress Test

An upright exercise stress test was performed in accordance with the standard Bruce protocol (Cowan et al., 1988), and athletes were encouraged to exercise to the point of achieving maximal age-predicted heart rate (maximal heart rate of 220–age). Continuous 12-lead ECG recordings were obtained throughout the test to look for episodes of polymorphic VT. ECG tracings were printed for the QT_c calculation at heart rate increments of 10 beats per minute up to a heart rate of 130 beats·min⁻¹ during exercise and at heart rate decrements of 10 beats·min⁻¹ from a heart rate of 130 beats·min⁻¹ to baseline heart rate during recovery. The QT_c was measured in lead V_3 during the exercise test because it usually has the largest T-wave amplitude (Wang et al., 1996).

Assessment of First-Degree Relatives

The first-degree relatives of the athletes with prolonged QT_c intervals (parents and siblings) were invited to undergo 12-lead ECGs to help identify evidence of familial disease.

Genetic Testing

All athletes with prolonged QT_c intervals were offered genetic testing for all genetic mutations commonly implicated in LQTS types 1 through 3 (KCNQ1, HERG, SCN5A). Genetic testing was performed following counseling and after obtaining informed consent. Mutations were identified using standard genetic tests (Curran et al., 1995; Schwartz et al., 1975; Wang et al., 1995).

Results

Of the 2,000 athletes, seven (six male and one female) had prolonged QT_c intervals amounting to a prevalence of 0.4%. The mean heart rate in these seven athletes was 58 beats·min⁻¹ (range was 47 to 68 beats per minute), and the QT_c ranged from 460 to 570 ms. Of the seven athletes, three had a baseline QT_c >500 ms (see figure 7.3). All seven athletes were asymptomatic; none were taking regular medications that could be associated with a prolonged QT_c interval; and none had a family history of congenital LQTS, premature SCD, unheralded syncope, or epilepsy. None of the athletes had sensorineuronal deafness (Moss, 1986; Schwartz, 1997). The characteristics of all athletes with prolonged QT_c interval are shown in table 7.2.

48-Hour ECG

All athletes with prolonged QT_c intervals successfully completed 48-hour ECG monitoring. None of the athletes showed evidence of polymorphic VT during the recording.

Exercise Stress Test

All athletes achieved at least 90% of their age-predicted heart rate during the test. None of the athletes developed episodes of polymorphic VT; however, two athletes exhibited prolongation of the QT_c interval during the initial stages of exercise and immediately postexercise (see figure 7.4 on p. 136). Both athletes had a baseline QT_c >500 ms.

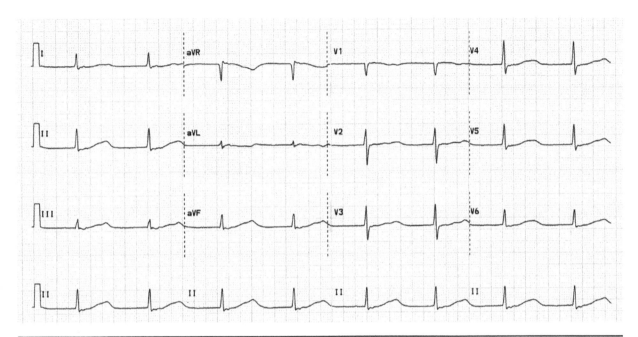

Figure 7.3 The 12-lead ECGs of three athletes exhibiting a QT$_c$ >500 ms.

Table 7.2 Characteristics of the Seven Athletes With Prolonged QT$_c$ Intervals

	Age	Gender	Sport	QT$_c$ interval	Positive ETT	Affected first-degree family members	Genetic testing
Athlete 1	17	Male	Swimming	532	Yes	Yes	Negative
Athlete 2	19	Male	Rugby	570	Yes	No	Negative
Athlete 3	16	Female	Swimming	515	No	Yes	Positive for LQT1
Athlete 4	15	Male	Football	460	No	No	Declined
Athlete 5	19	Male	Rugby	492	No	No	Declined
Athlete 6	15	Male	Tennis	474	No	No	Negative
Athlete 7	18	Male	Tennis	490	No	No	Negative

12-Lead ECG Screening of First-Degree Relatives

Both parents and all siblings of each of the seven athletes agreed to be evaluated with 12-lead electrocardiograms. One athlete had three siblings, two athletes had two siblings, and four athletes had one sibling. Two athletes had a first-degree relative with a long QT$_c$. In each case one parent and one sibling were affected (see figure 7.5 on p. 136). Both athletes had a baseline QT$_c$ >500 ms.

Genetic Testing

Of the seven athletes, two declined genetic testing after counseling. The results from the

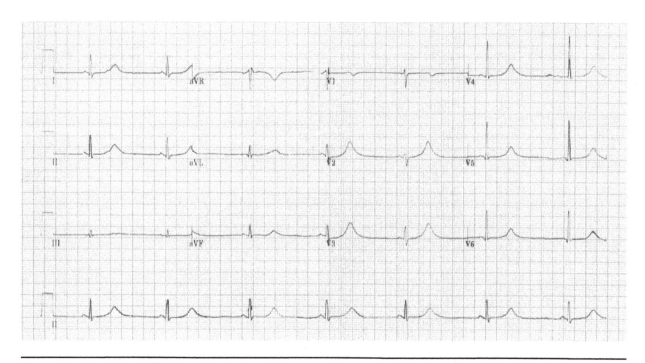

Figure 7.4 The 12-lead ECG of an athlete (athlete 2) demonstrating paradoxical prolongation of the QT$_c$ during the recovery phase of exercise.

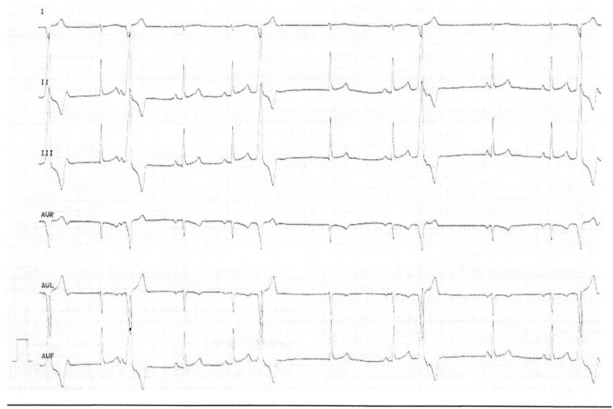

Figure 7.5 The 12-lead ECG of the brother of an athlete (athlete 3) with a long QT$_c$.

genetic tests took over four months to arrive for all athletes, and one athlete's results did not arrive until 12 months after submission. Only one of the five athletes (20%) who underwent genetic testing had a positive genetic diagnosis. The athlete in question had C-to-T nucleotide substitution at position c.691 of the KCNQ1 gene, resulting in an amino acid exchange of arginine (R) to cysteine (C) at codon 231 (p.R231C); baseline $QT_c > 500$ ms. A genetic diagnosis was not possible in the other four athletes after screening for all the known mutations capable of causing LQT1-3.

DISCUSSION

Prevalence of Isolated Long QT_c Interval on a 12-Lead ECG in Athletes

The diagnosis of congenital LQTS is based on the triad of prolonged QT_c interval on 12-lead ECG, unheralded syncope or polymorphic VT, and a family history of SCD or LQTS (Moss, 1986; Schwartz, 1997). The prevalence of LQTS based on this triad is between 1 in 2,500 to 1 in 10,000 (Quaglini et al., 2006; Zareba et al., 1998). However, genotype–phenotype correlation studies in patients with congenital LQTS have revealed that a significant number of gene-positive patients may manifest as a prolonged QT_c interval on the ECG in isolation.

The current study reveals that the prevalence of prolonged QT_c interval in athletes on the 12-lead ECG is 0.4%, which is higher than expected and not dissimilar to Mobitz type I second-degree AV block, wandering atrial pacemaker, and right bundle branch block, which are regarded as relatively rare but normal variants in athletes (Pelliccia et al., 2000). Indeed, based on the Italian preparticipation screening program comprised of over 34,000 athletes, the prevalence of prolonged QT_c interval is even higher and calculates to 0.69% (Nistri et al., 2003). The results of both studies suggest that the prevalence of LQTS is significantly higher than other cardiac disorders commonly implicated in

exercise-related sudden cardiac death in athletes, such as hypertrophic cardiomyopathy (Napolitano et al., 2005; Nistri et al., 2003). Considering the fact that up to 40% of people with LQTS will not be identified on a single 12-lead ECG, the prevalence of LQTS could be even higher (Funck-Brentano et al., 1993). The most feasible explanation for a higher prevalence of long QT in athletes is that those with ion channel disorders have preserved ventricular structure and function and have the potential to excel in sport in contrast with the vast majority of people with hypertrophic cardiomyopathy.

Significance of a Prolonged QT_c Interval in Athletes

Based on the observation that a prolonged QT_c interval may be the only phenotypic manifestation of LQTS, and the fact that some mutations, particularly those implicated in LQTS 1, are associated with adrenergic-mediated sudden cardiac death during physical activity, the identification of isolated prolongation of the QT_c (>440 ms) in an athlete is currently a recommendation for disqualification from competitive sport (Pelliccia et al., 2006; Sharma et al., 1999). However, current data on sudden death in young athletes (<35 years) indicate that SCD in the absence of a structural cardiac abnormality accounts for no more than 2 to 4% of cases in young athletes. In relation to the relatively high prevalence of a prolonged QT_c interval on the 12-lead ECG in athletes, the low death rates suggest that the vast majority of causal mutations may be relatively benign.

The actual significance of an isolated, prolonged QT_c interval in an athlete has never been studied. Although a prolonged QT_c interval >440 ms is usually considered to represent an ion channel disorder in the absence of causal drugs or electrolyte disturbances, it is recognized that athletes have a more prolonged QT_c interval than nonathletes. The exact mechanism for a prolonged

QT_c interval in athletes is uncertain but may represent the effect of delayed repolarization as a result of increased left ventricular mass, or the fact that Bazett's formula may not hold true in people with very slow heart rates (Priori et al., 2003). Conversely, a prolonged QT_c interval may be the only phenotypic manifestation of a potentially fatal ion channel disorder. In this case, abstinence from sport may be necessary to minimize the risk of sudden cardiac death. Clearly, the identification of a prolonged QT_c interval in the absence of symptoms or evidence of familial disease represents a clinical dilemma for the sport cardiologist.

The researchers studying these seven athletes aimed to eliminate the effects of physical training (increased left ventricular mass and slow heart rates) by reevaluating all athletes with a prolonged QT_c interval following a six-week period of detraining. Their study evaluated potentially affected athletes ($n = 7$) for the broader phenotypic features of the disorder and evidence of familial disease and showed that detailed evaluation with Holter monitoring, exercise stress test, and family screening proved useful in identifying additional phenotypic features of LQTS and provided diagnostic clarification in three of the seven athletes (43%).

Interestingly, all three of the seven athletes exhibited a baseline QT_c >500 ms and exhibited either paradoxical prolongation of the QT during exercise or had a first-degree relative with a prolonged QT_c interval. This observation suggests that the demonstration of a QT_c >500 ms in an athlete is indicative of unequivocal LQTS and warrants disqualification to minimize the risk of exercise-related sudden cardiac death particularly considering that a QT_c >500 ms is an independent risk factor for sudden death in probands with LQTS. In such cases, subsequent genetic testing may be useful in confirming the genotype and facilitating cascade screening if applicable.

The Gray Zone

In contrast to athletes with a baseline QT_c >500 ms, none of the athletes with a QT_c <500 ms had any features of congenital LQTS based on exercise testing and Holter monitoring, or any family members with a prolonged QT_c interval. The significance of an isolated, prolonged QT_c interval <500 ms, which accounted for 50% of the athletes in the study, remains unknown but may represent a relatively benign group for whom close monitoring may be more appropriate than disqualification in the absence a genetic diagnosis. Indeed, risk stratification in probands with LQTS suggests that males with QT_c <500 ms generally represent a low-risk group (Malik et al., 2002).

Genetic Testing

Following genetic counseling, only five athletes agreed to undertake genetic testing highlighting the difficulties in obtaining consent for genetic testing in highly trained athletes. A genetic diagnosis was not possible in a timely fashion and cost all five athletes their fitness and team selection while they abstained from sport. However, for the only athlete with confirmatory evidence of a disease-causing mutation, genetic testing proved invaluable because the mutation identified (KCNQI) is most commonly associated with exercise-related ventricular arrhythmias. The athlete had a baseline QT_c of >500 ms, the broader aspects of disease phenotype, and familial disease suggesting that a genetic test is probably unnecessary to aid the diagnosis of LQTS in athletes with a QT_c >500 ms. In the remaining four athletes, a genetic diagnosis could not be made despite extensive analysis suggesting that these athletes either had mutations that have not as yet been identified or did not have congenital LQTS.

Management

All athletes with a QT_c >500 ms exhibited other features of congenital LQTS to warrant disqualification from competitive sport. However, the remaining four athletes with QT_c >500 ms did not demonstrate the broader phenotype of the syndrome or evidence of disease in first-degree relatives. In these athletes the absence of objective features of the disorder made it difficult to enforce abstinence from participation. Clinicians made a clinical decision to allow the athletes to continue to participate in competitive sport, contrary to published guidelines (Pelliccia et al., 2006). This is because there was no mandatory obligation in the UK at the time to disqualify an athlete with a prolonged QT_c from competitive sport, unless results from genetic testing suggested otherwise. Enforcing abstinence from sport while awaiting genotyping would have resulted in unacceptable delays for the athletes and would not have been useful in the diagnosis. Fortunately, all four athletes remained well and asymptomatic after a mean follow-up of almost three years.

TAKE-HOME MESSAGE

The prevalence of prolonged QT_c interval in asymptomatic athletes is considerably higher than expected. In asymptomatic athletes without a family history of LQTS, the identification of a QT_c >500 ms is highly suggestive of LQTS and warrants disqualification from competitive sport. The significance of a QT_c <500 ms in asymptomatic athletes is less clear; however, this study suggests that in this group of athletes, the absence of other phenotypic features of LQTS on Holter monitoring and exercise stress testing results or indications of familial disease on 12-lead ECGs should not necessarily result in disqualification from sport.

S. Basavarajaiah, M. Wilson, G. Whyte, A. Shah, E. Behr, and S. Sharma, "Prevalence and significance of an isolated long QT interval in elite athletes," *European Heart Journal* 2007, 28(23): 2944-2949, by permission of Oxford University Press.

CASE 4 STUDY

Differentiation of RVOT-VT and ARVC in an Elite Athlete

BACKGROUND

In October 2005, a 25-year-old Afro-Caribbean female international track and field sprinter attended the CRY Centre for Sports Cardiology at the British Olympic Medical Institute following the presentation of symptoms including chest tightness, inappropriate dyspnea accompanied by an inappropriate tachycardia (>220 beats·min^{-1} measured by portable telemetry system) during exercise, and underperformance. There was no family history of note. Physical examination demonstrated resting bradycardia with normal heart sounds, blood pressure, and venous pressure and no ankle edema. Resting 12-lead ECG demonstrated sinus rhythm with frequent monomorphic extra systoles (LBBB

morphology) and runs of monomorphic VT at a rate of 140 beats·min^{-1} (see figure 7.6 on p. 140). The 24-hour ECG demonstrated similar findings with >20% of the total time spent in VT (see figure 7.7 on p. 141). Integrated cardiopulmonary stress test demonstrated a normal $\dot{V}O_2$max, heart rate, and blood response to exercise. Of note, during exercise the ventricular ectopics disappeared at 140 beats·min^{-1} and reappeared during cool-down at a similar heart rate.

Echocardiography demonstrated normal left ventricular (LV) wall thickness and LV systolic and diastolic function. Valves and color flow were normal. The right ventricle (RV) appeared dilated and following contrast enhancement demonstrated hypokinesia,

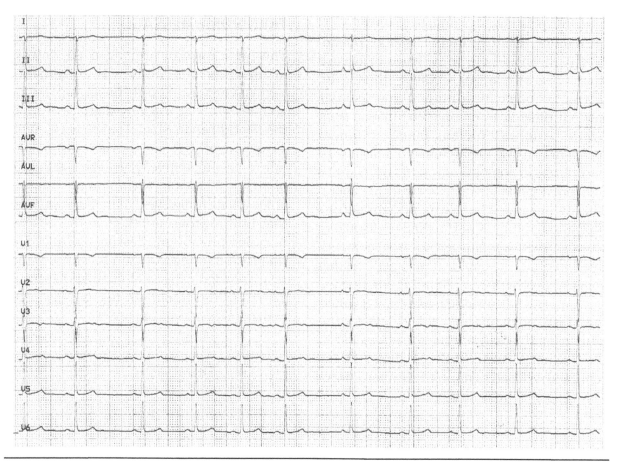

Figure 7.6   Initial presenting resting ECG demonstrating trigeminy (right ventricular origin).

especially in the apex, with slow contrast clearance. Cardiac magnetic resonance with late gadolinium enhancement demonstrated normal cardiac morphology with a mildly impaired LV and RV function. There was no evidence of active inflammation or fibrosis. Cardiac catheterization demonstrated an abnormal RV pressure tracing with mild LV systolic functional impairment more noticeable in the basal and apical regions. RV angiography demonstrated a dilated RV with moderate to poor contraction. At this point the athlete was instructed to cease training and commenced a course of medication (carvedilol and ramipril).

At a one-month follow-up the athlete reported no change in symptoms and unaltered findings on resting 12-lead and 24-hour ECG. The athlete was referred for radiofrequency ablation. Using pace mapping,

the ectopic focus was successfully ablated. Follow-up examination demonstrated normal resting 12-lead (see figure 7.8) and 24-hour ECG. Echocardiography demonstrated normal LV dimensions and function and a minimal RV dilation with well-preserved RV systolic function. The athlete was cleared for return to exercise, and at a three-month follow-up she was entirely asymptomatic with normal findings. At 12- and 24-month follow-ups the athlete was well and in full training and competition. Informed written consent was obtained to use the clinical data in this case report.

DISCUSSION

An increased incidence of supraventricular, profound bradyarrhythmia, and complex ventricular arrhythmias is observed in athletes. Differentiating physiological from pathological substrates for these arrhythmias

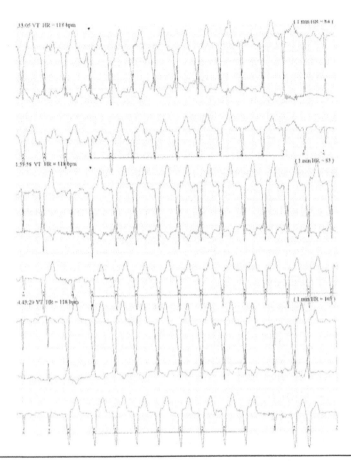

Figure 7.7 Twenty-four-hour ECG demonstrating multiple episodes of non-sustained VT.

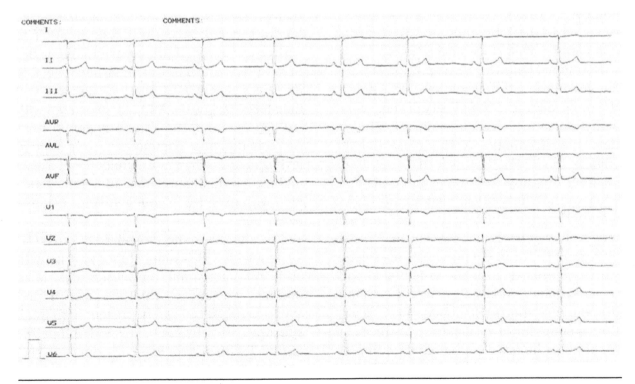

Figure 7.8 Twelve-lead ECG at one-month follow-up.

is important given the potential for adverse events. A recent study reported findings from 46 well-trained endurance athletes presenting with ventricular arrhythmia (VA). Eighty percent of the arrhythmias had a left bundle branch morphology, and a right ventricular arrhythmogenic involvement was manifest in 59% of athletes and suggestive in 30%. The prognosis for these athletes was poor with a sudden death incidence of 25% (all cyclists) at a 4.7-year follow-up (Heidbuchel et al., 2003). These findings are in contrast to those of Biffi and colleagues (2002), who reported a very low incidence of death (1 in 355 athletes over an eight-year follow-up) in athletes presenting with frequent and complex ventricular tachyarrhythmias.

Care is warranted in comparing these studies because a number of key differences exist. Of the 355 athletes in the Biffi study, only 38 (11%) presented with nonsustained VT (2) compared to 100% of the athletes in the Heidbuchel (2003) study. Furthermore, all 38 athletes in the Biffi study were disqualified from competition, whereas those in the Heidbuchel study continued to train and compete. Of note, the death of one athlete reported in the Biffi study occurred during competition against medical advice.

Detraining following the cessation of training and competition leads to a reduction in the prevalence and complexity of ventricular arrhythmias, which underscores the importance of restricting exercise in these athletes (Biffi et al., 2004). The majority of athletes in the Biffi study were team game players, whereas those in the Heidbuchel study (2003) were endurance athletes. The significance of mode of exercise is unclear but may be an important factor in the prognosis of athletes with arrhythmias, as is the potential impact of performance-enhancing drug use common in endurance sports (Heidbuchel et al., 2003). Both studies, however, demonstrate the need for comprehensive follow-up of complex

ventricular arrhythmias (Biffi et al., 2002; Heidbuchel et al., 2003). It is estimated that <10% of patients who present with ventricular tachycardia (VT) have no obvious structural disease (Biffi et al., 2002; Heidbuchel et al., 2003). Types of VT that occur in the absence of structural heart disease include RV monomorphic extrasystoles, RV outflow tract VT (RVOT-VT), left ventricular outflow tract VT (LVOT-VT), idiopathic LV tachycardia (ILVT), idiopathic propanolol-sensitive (automatic) VT (IPVT), catecholaminergic polymorphic VT (CPVT), Brugada syndrome, and long QT syndrome (LQTS). In the case of idiopathic VT from the RVOT and LVOT, the arrhythmia is monomorphic and generally not familial.

RV monomorphic extrasystoles and RVOT-VT appear to be on a continuum of the same process. In both disorders the resting ECG has no identifiable abnormalities, and the echocardiogram and coronary angiography are normal. The MRI may show abnormalities in up to 70% of patients, including focal thinning, diminished wall thickening, and abnormal wall motion (Globits et al., 1997).

The prognosis for RVOT-VT is generally benign. However, differentiating idiopathic VT from occult structural disease such as arrhythmogenic right ventricular cardiomyopathy (ARVC) is critical because this diagnosis affects both ablation outcomes and long-term prognosis; ARVC is associated with sudden death. In the Veneto region of Italy, ARVC is the most common cause of sudden arrhythmic death in those under the age of 35 and the predominant cause of death associated with exercise in young athletes (Thiene et al., 1999).

The characteristic morphology of RVOT-VT is a wide QRS complex tachycardia with an LBBB pattern and an inferior axis. Most patients present with palpitations or pre-syncope (50%) but rarely present with frank syncope (Srivathsan et al., 2005). Idiopathic LBBB VT with inferior axis can originate along the right ventricular outflow tract (RVOT-VT) and into the pulmonary artery (Thiene et al., 1988). Two phenotypic forms of RVOT-VT exist: nonsustained, repetitive, monomorphic VT and paroxysmal, exercise-induced, sustained VT. The extrasystoles observed at rest may diminish or disappear during exercise in RV monomorphic extrasystoles (Leclercq et al., 1981). In contrast, exercise (or emotional stress) may precipitate VT in RVOT-VT. Given the continuum of RV monomorphic extrasystoles and RVOT-VT, however, extrasystoles may disappear during exercise in RVOT-VT, as was the case in this case study.

Intracellular calcium overload appears to be the principal mechanism of RVOT-VT. Cyclic adenosine monophosphate (cAMP) has a substantial role in regulating intracellular calcium; thus, when the concentration of cAMP is increased, intracellular calcium levels are high. Acute termination of RVOT-VT can be achieved by vagal maneuver or intravenous adenosine (adenosine lowers cAMP concentration). Adenosine is considered to be a specific mechanistic probe with regard to its effects on VT. In the absence of cAMP stimulation, adenosine has no effect on the slow inward calcium current or the transient sodium current, but in the presence of catecholamine stimulation it attenuates these currents, accounting for its termination of VT as a result of triggered activity and cAMP-mediated delayed afterdepolarizations. This effect appears to be specific because adenosine has no effect on reentrant VT related to structural heart disease, regardless of whether catecholamines are required for facilitation. Although RVOT-VT is adenosine sensitive, the majority of patients presenting to the emergency room would not be treated with adenosine because this would go against accepted guidelines for the treatment of stable VT. Acute termination of RVOT-VT with adenosine requires specialist consultation.

Treatment of RVOT-VT includes beta-blockers that inhibit adenylate cyclase leading to a decrease of cAMP, and calcium channel blockers that decrease the concentration of intracellular calcium. Unfortunately, pharmacotherapy has only a 25 to 50% efficacy rate (Buxton et al., 1983), a finding supported by this case study. In contrast, the preferred course of treatment is radiofrequency ablation with a 90 to 95% cure rate and ≤1% occurrence of major complication (Klein et al., 1992). Furthermore, pharmacotherapy may contravene the World Anti-Doping Agency (WADA) regulations and is therefore often contraindicated in elite athletes.

Arrhythmogenic right ventricular cardiomyopathy (ARVC) is a heart muscle disorder characterized by structural and functional abnormalities of the right ventricle due to a fibrofatty replacement of the myocardium. These fibrofatty infiltrates predispose the heart to reentrant ventricular arrhythmias (Sen-Chowdhry et al., 2004). These arrhythmias are typically induced by adrenergic stimulation such as catecholamine infusion or physical exertion (Gemayal et al., 2001).

ARVC is a disease of myocardial cell-to-cell adhesion and is typically inherited as an autosomal dominant trait with variable penetrance and incomplete expression. The genes responsible for ARVC have not been identified, but seven loci have been mapped to four chromosomes (1, 3, 10, 14) (Gemayal, et al., 2001). These disease-causing loci are associated with desmosomal proteins including plakoglobin, desmoplakin, plakophilin, and desmoglein (Tsatsopoulou et al., 2006).

The genetic etiology of ARVC, coupled with the high incidence of sudden death as the first clinical manifestation, highlights the importance of screening first-degree relatives to identify affected individuals and confirm diagnosis. In this case study the athlete's first-degree relatives declined screening; therefore, examination of familial cardiovascular disease was not possible. The reluctance of first-degree relatives to undertake screening is not uncommon and, combined with the present lack of conclusive genetic screening, highlights the importance of noninvasive diagnostic techniques in the diagnosis of familial cardiovascular disease.

The natural history of ARVC is considered to include four phases (Sen-Chowdhry et al., 2004). The early clinical manifestation is RV-originating tachyarrhythmia during adolescence and early adulthood after a latent (concealed) phase demonstrating ECG abnormalities. Because sudden death is often the first symptom during this phase (Heidbuchel et al., 2003), early noninvasive diagnosis is important. Clinical diagnosis of ARVC is challenging owing to the nonspecific nature of associated findings and the absence of a single diagnostic test (Sen-Chowdhry et al., 2004). As a result, major and minor criteria relating to structural, histological, electrocardiographic, arrhythmic, and genetic features of the disease have been established (Richter and Hohnloser, 2006). The presence of two major, one major and two minor, or four minor criteria in different categories is considered diagnostic. Although these criteria are highly specific, they lack sensitivity for early disease (Sen-Chowdhry et al., 2004).

In ARVC the resting 12-lead ECG typically demonstrates T-wave inversions in the right precordial leads, which is a common finding in athletes (Sharma et al., 1999). Epsilon waves (prominent in V_1 and V_2) are considered the electrocardiographic hallmark of ARVC; however, they are not always present on standard 12-lead ECG (Richter and Hohnloser, 2006). The 12-lead ECG is normal in patients with RVOT-VT (as observed in this case study); however, this is also true for many patients with early ARVC (Sen-Chowdhry et al., 2004). Although VT in ARVC may have morphological features similar to RVOT-VT (LBBB with inferior axis), (Gemayel et al., 2001) the morphology and duration of the QRS during VT may be useful in the differential diagnosis. The mean QRS duration during VT is greater in all leads in ARVC patients compared with RVOT-VT. These differences are likely associated with absence of slowly conducting tissue in RVOT-VT. These findings are supported by endocardial pace mapping studies in which a septal or near-septal origin, uncommon in ARVC, resulted in shorter QRS durations (Ainsworth et al., 2006). Therefore, a QRS duration ≥120 ms in lead I and a QRS axis <30° are predictive of ARVC (Ainsworth et al., 2006).

The differentiation of RVOT-VT and ARVC with ECG alone, particularly during the concealed phase, may be difficult and should be viewed with caution. Electrophysiological study may assist in the differential diagnosis because RVOT-VT patients appear to have a high prevalence of isoproterenol-provocable and programmed stimulation (atrial or ventricular pacing) VT (Biffi et al., 2004).

Although ARVC is primarily associated with the right ventricle, left ventricular involvement may occur with disease progression (Gemayel et al., 2001; Hirimoto et al., 2000). Therefore, careful examination of the left ventricle is warranted (Sen-Chowdhry et al., 2004). Of note, LV involvement is uncommon in the early concealed phase of ARVC. In this case study hypokinesis and impaired systolic function of the right ventricle concomitant to impaired left ventricular dysfunction was observed. The interpretation of this finding should be viewed with caution, however, because benign long-standing sustained or nonsustained VT leads to myocardial hypokinesis, which may affect both the right and left ventricle. Ablation of the ectopic focus can reverse this hypokinesis and impaired function and return the right and left ventricles to normal function (Whyte et al., 2007). This finding is supported by Yarlagadda and colleagues (2005), who reported the reversal of unexplained (idiopathic) cardiomyopathy in patients with repetitive monomorphic ventricular ectopy originating from the RVOT. Although the mechanisms responsible for the observed dysfunction are not known, the authors noted a normalization of ventricular function following ablation.

Visualization of the right ventricle by echocardiography is often unsatisfactory. Contrast echocardiography significantly enhances the imaging of the RV; however, cardiac magnetic resonance is considered the gold standard for noninvasive tissue characterization and is indicated in the initial evaluation of those with suspected ARVC. Care is warranted in the identification of fibrofatty replacement because fat deposits are observed in normal hearts (Sen-Chowdhry et al., 2004). Furthermore, the fibrofatty replacement observed in ARVC is often microscopic rather than overt, leading to false negative findings. Although impaired right and left ventricular function was observed in this case study, the MRI failed to demonstrate a pattern typical of ARVC. Nevertheless, the early concealed phase could not be excluded.

TAKE-HOME MESSAGE

An increased incidence of supraventricular, profound bradyarrhythmia and complex ventricular arrhythmias is observed in athletes.

Differentiating physiological from pathological substrates for these arrhythmias is important given the potential for adverse event. The differentiation of RVOT-VT and the early concealed phase of ARVC is clinically challenging. Sudden death is often the first symptom during this concealed phase and highlights the importance of early diagnosis. Extensive cardiac investigation may be unremarkable in the concealed phase of ARVC. Care is warranted in the interpretation of cardiac investigation in athletes with RVOT-VT. ECG anomalies associated with the athlete's heart may mimic those observed in ARVC. Furthermore, impaired cardiac function may be associated with long-standing VT in the absence of underlying disease. Radiofrequency ablation may be a valuable tool in the differential diagnosis because this technique is highly effective in the treatment of RVOT-VT and of limited value in ARVC. Continued follow-up evaluation is important for the confirmation of disease status following the diagnosis of idiopathic RVOT-VT.

Adapted, by permission, from G. Whyte et al., 2008, "Differentiation of RVOT-VT and ARVC in an elite athlete," *Medicine and Science in Sports and Exercise* 40(8):1357-1361.

CASE **5** STUDY

Treat the Patient Not the Blood Test: The Implications of an Elevation in Cardiac Troponin Following Prolonged Endurance Exercise

BACKGROUND

In April 2006, a 43-year-old Caucasian male club-level triathlete presented to the CRY Centre for Sports Cardiology at the British Olympic Medical Centre for a follow-up consultation. In March 2006, following completion of an Ironman triathlon (3.8 km swim, 180 km bike, 42 km run) in a personal best time, the athlete received a massage during

which he experienced presyncopal symptoms. The athlete received fluid resuscitation (500 ml, 1% saline) and recovered well. Following a long period of standing in the heat, the athlete reported presyncopal symptoms and collapsed, without syncope. Blood pressure was measured at 75/40 mmHg. Following further fluid resuscitation (500 ml, 1% saline), the athlete was referred to the cardiology department of a local hospital. He was conscious and alert with normal core temperature, blood glucose, and plasma sodium, and was otherwise asymptomatic.

On admission to the hospital, the athlete's ECG demonstrated minor ST-segment elevation in anterior lateral and inferior leads (see figure 7.9), and his cardiac troponin-I (cTnI) was mildly elevated at 0.06 U.l⁻¹. Echocardiography was normal with mild, concentric left ventricular hypertrophy. The athlete underwent diagnostic cardiac catheterization, which demonstrated normal coronary arteries. Based on the mildly elevated cTnI and minor ST-segment elevation, the athlete was diagnosed with mild myopericarditis and advised to avoid exercise for six weeks. On follow-up two weeks after being discharged from the hospital, the resting 12-lead ECG demonstrated similar findings of marginal ST-segment elevation in anterior lateral and inferior leads (see figure 7.10). Echocardiography demonstrated mild concentric left ventricular hypertrophy (12 mm) in the presence of normal diastolic and systolic function. There was no evidence of myocarditis or pericarditis, and the athlete

was immediately cleared for training and competition.

DISCUSSION

Exercise-associated collapse is common following prolonged endurance exercise (O'Connor et al., 2003) and is usually benign. However, in some circumstances it can be severe and life threatening. Benign causes of collapse include exhaustion, postural hypotension, dehydration, and leg cramps. Serious causes include hyponatremia (low blood sodium), heatstroke, hypoglycemia, hypothermia, cardiac arrest, and other medical conditions (Sallis, 2004). In the presence of a conscious, alert athlete presenting with normal blood glucose and plasma sodium, and normal core temperature, as observed in the athlete in this case study, the likely cause of collapse is a vasovagal response leading to postural hypotension that may be exacerbated by dehydration.

Although the health benefits of regular, moderate-intensity physical activity are well recognized, a growing body of evidence demonstrates an acute reduction in cardiac function and the presence of humeral markers

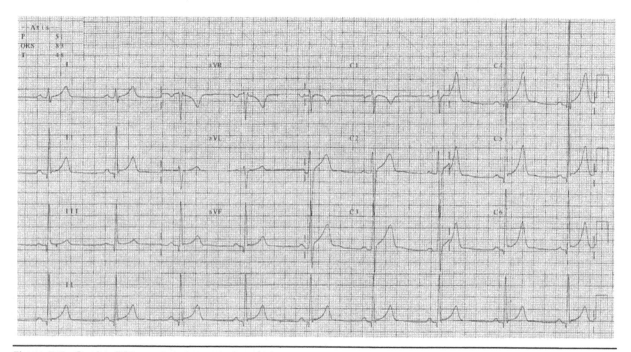

Figure 7.9   Resting ECG on admission to the hospital.

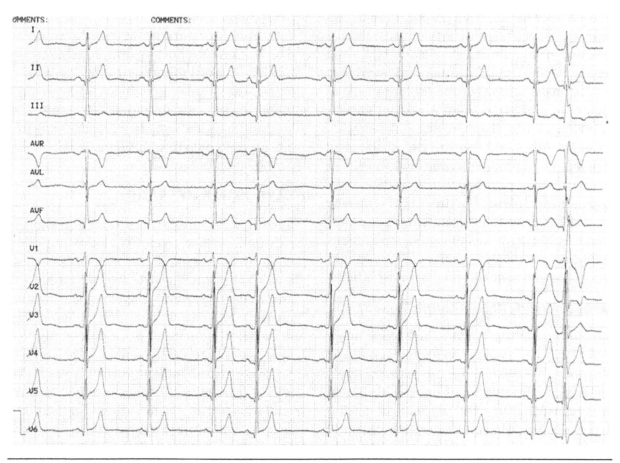

COMMENTS: COMMENTS:

I

II

III

AVR

AVL

AVF

V1

V2

V3

V4

V5

V6

Figure 7.10   Resting ECG on presentation at follow-up following hospitalization.

of cardiac myocyte damage (cTnT and cTnI) that may be above acute myocardial infarction cutoff levels following ultra-endurance exercise (Dawson et al., 2003). Among this work are a number of studies examining the Ironman triathlon (Douglas et al., 1987; Whyte et al., 2000).

The prevalence of postexercise elevations in cTnT, cTnI, or both, are not fully known. However, in a recent study, 78% of runners investigated following a marathon presented evidence of minor cardiomyocyte damage (Shave et al., 2005). Previous studies have suggested that these changes are likely to be physiological in nature due to the athletes' rapid (<24 hours) return to baseline values and of no clinical significance. Indeed, the latest work from one lab demonstrates a peak in cTnT at three hours postexercise with a subsequent fall to baseline levels at 24 hours

postexercise (unpublished findings). These findings are in direct contrast to the kinetics of cardiac troponin release observed following myocardial infarction, in which an initial release of unbound troponin is followed by a continued release of the structurally bound troponin as it degrades, resulting in a sustained elevation in circulating troponin for many days following infarction (Newby, 2004).

The level of cTnI observed in the athlete in this case study is similar to that previously described in athletes following ultra-endurance exercise and likely represented a physiological response to prolonged arduous exercise. The use of one-off measures of cardiac troponins should be viewed with caution following prolonged exercise. Serial measures would help eliminate myocardial infarction as a cause of the elevated cardiac troponins.

A number of ECG changes observed in athletes are associated with physiological anomalies and of no clinical significance (Sharma et al., 1999). The diffuse minor ST-segment elevations observed in the athlete's ECG in this case study represent a normal variant as evidenced by their presence at follow-up examination. Furthermore, the absence of echocardiographic evidence of myocardial or pericardial injury supports the evidence of a psychological cause. The use of ECG screening of athletes prior to ultra-endurance competition may aid in the identification of pathology following prolonged exercise and avoid misdiagnosis and mismanagement of the athlete.

What Is Already Known on This Topic

An altered cardiac function and blood markers of cardiac damage following prolonged arduous exercise have been extensively reported in the literature since the late 1990s. It is postulated that the presence of cardiac troponins following prolonged exercise and a rapid return to baseline (<24 hours), although pathognomonic of cardiac damage, are physiological in nature and of no long-term significance.

What This Study Adds

This study highlights the difficulties faced by medical support teams and hospital medical staff when dealing with athletes following prolonged arduous exercise. Despite the extensive literature reporting elevated cardiac troponins following prolonged exercise, this case study demonstrates the lack of awareness of this phenomena and the often inappropriate care provided as a result. This study offers advice, guidance, and supporting literature to help practitioners deal with this problematic phenomenon.

TAKE-HOME MESSAGE

Care is warranted in the interpretation of elevated cardiac troponins following ultra-endurance exercise. Serial measures of cTnT and cTnI following exercise may help differentiate underlying physiological mechanisms from pathological mechanisms. The use of ECG screening prior to competition may assist in the management of the athlete postexercise. Misdiagnosis of myocardial injury following ultra-endurance exercise and subsequent mismanagement, including hospitalization and invasive intervention, can be expensive and psychologically damaging to the athlete. Diagnosis of myocardial injury following prolonged exercise should be made based on all available information and not blood tests alone.

Adapted from *British Journal of Sports Medicine,* "Treat the patient not the blood test: The implications of an elevation in cardiac troponin following prolonged exercise," G. Whyte et al., 1(9): 613-615. © 2007 with permission from BMJ Publishing Group Ltd.

CASE **6** STUDY

Acute Myocardial Infarction in the Presence of Normal Coronaries and the Absence of Risk Factors in a Young, Lifelong Regular Exerciser

BACKGROUND

In September 2007, a 46-year-old male of Anglo-Indian descent developed chest pain following a regular exercise session at his local gym. The man was a lifelong regular exerciser with no cardiovascular risk factors and no family history of note and was in good health at the time of the incident. The pain was initially thought to be indigestion but worsened over three hours, at which point he presented to the emergency room of his local hospital. On admission, an ECG confirmed inferolateral ST-segment elevation. Creatine kinase and cardiac troponin I concentrations were 1025 μ·L⁻¹ and 29.5 μ·L⁻¹, respectively. On transfer to a tertiary hospital for primary percutaneous intervention (PCI), he suffered a VF arrest that was cardioverted to sinus rhythm with a single 360 J shock. Coronary angiography demonstrated no obstructive disease in his left coronary artery. His right coronary artery was a large dominant vessel that was occluded in the midcourse with extensive intraluminal thrombus. Multiple aspirations were performed with an Export Thrombus Aspiration Catheter. This restored flow distally; however, significant thrombus remained in the proximal and distal vessel. A ReoPro bolus was administered concomitant to multiple doses of adenosine and GTN spray. Further aspirations were performed using ThromCat with further improvement in flow.

Because of the heavy burden of thrombus, it was decided not to proceed with balloon angioplasty or stent deployment. The patient was commenced on a 24-hour infusion of ReoPro and transferred to the hospital. Two days later he was restudied and an intravascular ultrasound was performed. This confirmed the presence of an organized thrombus in the mid and distal right coronary artery, although neither balloon angioplasty nor stent deployment was considered necessary given the good flow. Echocardiography on the same day demonstrated good LV function with no obvious hypokinesis of the patient's inferior wall, and RV function was normal. The patient was discharged three days following acute myocardial infarction.

At the one-month follow-up, technetium scintigraphy demonstrated a fixed perfusion defect in the man's inferior wall with preserved LV systolic function. A bubble contrast echocardiogram was performed using agitated saline at two months post-AMI, which demonstrated proximal septal wall thinning with marked dyskinesis and but no evidence of patent foramen ovale (PFO). Cardiac magnetic resonance (CMR) with adenosine stress perfusion study performed two months after post-AMI demonstrated normal LV and RV

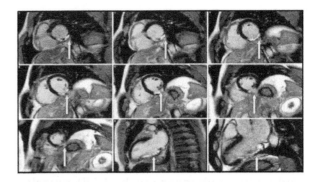

Figure 7.11 Late gadolinium-enhanced CMR demonstrating inferior wall infarction extending from the base to the apex of the LV (white areas indicated by arrows).

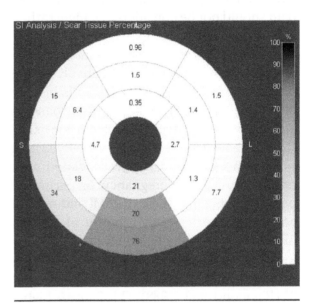

Figure 7.12 Bull's eye plot representing the distribution of scar tissue in the left ventricle. Plot demonstrates 21.7g of scar tissue (17% of total LV mass) concentrated in the inferior LV wall.

function in the presence of an inferior wall infarction extending from the base to the apex affecting 17% (21.7 g) of the total LV mass (128 g) (see figures 7.11 and 7.12). At present the man remains in good health and has recommenced regular moderate-intensity exercise avoiding excessive rises in systemic blood pressure.

DISCUSSION

Around 6% of patients suffering an AMI have normal coronary arteries. The mechanisms responsible are not fully known but include a hypercoagulable state, coronary endothelial dysfunction, aortic dissection, inflammation, coronary thrombosis, aortic wall stiffening, cocaine abuse, carbon monoxide poisoning, and paradoxical embolism. In this case a man who participated in lifelong regular exercise without risk factors for cardiovascular disease suffered an AMI with normal coronaries. Despite normal cardiac function on left ventriculography and echocardiography, late gadolinium enhancement by CMR revealed significant cardiac necrosis. The long-term prognosis is favorable with low rates of coronary morbidity and mortality. Acute chest pain should not be considered benign; rather, it warrants medical investigation.

AMI in the presence of normal coronary arteries typically occurs in relatively young people (<40 years) with no previous history of chest pain, no hemostatic disturbances, and no risk factors other than smoking (Chandrasekaran and Kurbaan, 2002). In women, hypercoagulability associated with oral contraception, pregnancy, or peripartal period has been linked with AMI and normal coronary arteries (Raymond et al., 1988). Around 6% of patients suffering an AMI are found to have normal coronary arteries. This figure rises to 10% in patients under the age of 35 (Raymond et al., 1988). This figure is highly variable in the literature because of varied definitions of *normal* and the level of investigation undertaken to identify the presence of disease. In a large study examining 1,150 patients presenting with AMI, the incidence of angiographically normal coronary arteries was 2.6%. Following follow-up evaluation, the incidence of truly normal coronaries fell to 0.7% (Widimsky et al., 2006). These findings support the role for extensive evaluation of patients presenting with AMI and angiographically normal coronaries to fully establish the etiology, level of damage, and risk stratification.

The mechanisms underlying AMI with normal coronaries are not fully understood. A number of mechanisms have been postulated in isolation or combined, including hypercoagulable state, coronary endothelial dysfunction (coronary vasospasm), aortic dissection, inflammation (viral myocarditis, autoimmune vasculitis), coronary thrombosis (in situ formation or embolus), aortic wall stiffening, cocaine abuse, and carbon monoxide poisoning (Chandrasekaran and Kurbaan, 2002; Kalaga et al., 2007; Robisek, 2002). Recent reports suggest a possible role for paradoxical embolism as a result of venous-to-arterial circulation shunts caused by patent foramen ovale (Sastry et al., 2006).

Despite the absence of overt coronary artery disease (CAD), treatment assumes CAD as the most likely etiology (Chandrasekaran and Kurbaan, 2002). Indeed, it has been suggested that endothelial dysfunction, possibly superimposed to nonangiographically evident atherosclerosis, is likely to be the underlying common feature predisposing to the acute event (Kardasz and De Caterina, 2007). The long-term prognosis for patients suffering an AMI with normal coronaries is favorable with low rates of coronary morbidity and mortality. Recurrent infarction, postinfarction angina, heart failure, and sudden cardiac death are rare. Traditionally, it is believed that long-term survival is principally associated with residual left ventricular function, which is usually good (Tun and Khan, 2001). Recent findings from CMR (cardiovascular magnetic resonance) suggest that the volume of necrosis is a valuable measure in risk stratification (Schmidt et al., 2007). Although normal epicardial arteries that display normal intravascular ultrasound and normal LV function are important prognostically, they fail to elucidate the level of cardiac damage. CMR can highlight significant regions of infarct and may confer additional prognostic information and inform exercise prescription.

What Is Already Known on This Topic

- Angina-like chest pain in the absence of identifiable coronary artery disease is a common syndrome.
- Six percent of AMI patients present with normal coronaries at angiography.

- The characteristics of patients presenting with AMI and normal coronaries include a relatively young age (<40 years), no previous history of chest pain, no hemostatic disturbances, and no risk factors other than smoking.

What This Study Adds

- AMI with normal coronaries can occur in relatively young lifelong exercisers in the absence of cardiac risk factors and family history.

- "Time Is Muscle"—Acute chest pain should not be considered benign; it warrants rapid medical investigation.

- The presence of cardiac troponins is pathognomonic for cardiac damage. CMR is a valuable tool in identifying the extent of myocardial damage and assists in the short- and long-term management of the patient.

TAKE-HOME MESSAGE

Acute chest pain in a young, previously asymptomatic person without risk factors for CAD should not be assumed to be benign. Acute myocardial infarction in the presence of normal coronary arteries is observed in 6% of patients undergoing coronary angiography. The heterogenic nature of possible underlying mechanisms for AMI with normal coronaries warrants extensive evaluation to establish etiology and the level of myocardial damage to optimize treatment and assist in risk stratification.

Adapted from *British Medical Journal Case Reports,* "Acute myocardial infarction in the presence of normal coronaries and the absence of risk factors in a young, lifelong regular exerciser," G. Whyte, R. Godfrey, M. Wilson, J. Buckley, and S. Sharma, © 2009 with permission from BMJ Publishing Group Ltd.

References

Chapter 5

American College of Sports Medicine (ACSM). *ACSM's Guidelines for Exercise Testing and Prescription.* 5th ed. Media, PA: Williams & Wilkins; 2000.

Gibbons, R.J., Balady, G.J., Bricker, J.T., Chaitman, B.R., Fletcher, G.F., Froelicher, V.F., Mark, D.B., McCallister, B.D., Mooss, A.N., O'Reilly, M.G., Winters, W.L. ACC/AHA 2002 Guideline Update for Exercise Testing: A Report of the ACC/AHA Task Force on Practice Guidelines (Committee on Exercise Testing). American College of Cardiology. Available at: www.acc.org/clinical/guidelines/exercise/dirIndex.htm. 2002.

Chapter 6

Basavarajaiah, S., Boraita, A., Whyte, G., Wilson, M., Carby, L., Shah, A., Sharma, S. Ethnic differences in left ventricular remodeling in highly-trained athletes relevance to differentiating physiologic left ventricular hypertrophy from hypertrophic cardiomyopathy. *J Am Coll Cardiol.* 2008;51(23):2256-2262.

Basavarajaiah, S., Makan, J., Reza, S., Whyte, G., Gati, S., Sharma, S. Physiological upper limits of left atrial diameter in highly trained adolescent athletes. *JACC* 2006;47:2341-2342.

Batt, M.E., Jaques, R., Stone, M. Preparticipation examination (screening): Practical issues as determined by sport: a United Kingdom perspective. *Clin J Sport Med.* 2004;14(3):178-182.

Benefits and risks associated with exercise. In *American College of Sports Medicine Guideline for Exercise Testing and Prescription.* pp. 3-11. Baltimore: Williams & Wilkins; 1995.

Corrado, D., Basso, C., Schiavon, M., Thiene, G. Screening for hypertrophic cardiomyopathy in young athletes. *NEJM.* 1998;339:364-369.

Green, D. Maiorana, A., O'Driscoll, G., Taylor, R. Effect of exercise training on endothelium-derived nitric oxide function in humans. *J Physiol.* 2004;156:1-25.

Krediet, P., van Dijk, N., Linzer, M., van Lieshout, J., Wieling, W. Management of vasovagal syncope: Controlling or aborting faints by leg crossing and muscle tensing. *Circulation.* 2002;106:1684-1689.

Maron, B., Gohman, T., Kyle, S., Estes, N., Link, M. Clinical profile and spectrum of commotio cordis. *JAMA.* 2002;287:1142-1146.

Maron, B.J. The young competitive athlete with cardiovascular abnormalities: Causes of sudden death, detection by preparticipation screening, and standards for disqualification. *Cardiac Electrophysiology Review.* 2002;6:100-103.

Maron, B.J., Shirani, J., Poliac, L.C., Mathenge, R., Roberts, W.C., Mueller, F.O. Sudden death in young competitive athletes. Clinical, demographic and pathological profiles. *JAMA.* 1996;276:199-204.

Naylor, L., George, K., O'Driscoll, G., Green, D. The athlete's heart: A contemporary appraisal of the 'Morganroth Hypothesis'. *Sports Med.* 2008;38:1-21.

Papadakis, M., Whyte, G., Sharma, S. Preparticipation screening for cardiovascular abnormalities in young competitive athletes. *Br Med J.* 337:a1596. doi:10.1136/bmj.a1596. 2008.

Pelliccia, A., Maron, B.J., Spataro, A., Proschan, M., Spirito, P. The upper limits of physiologic cardiac hypertrophy in highly trained athletes. *NEJM.* 1991;324:295-301.

Privett, S., George, K. Middleton, N., Whyte, G., Cable, N.T. The effect of prolonged endurance exercise upon blood pressure regulation during a post-exercise orthostatic challenge. *Br J Sports Med.* 2010 Apr 19.

Seto, C.K. Preparticipation cardiovascular screening. *Clin Sports Med.* 2003:23-35.

Sharma, S., Elliott, P.M., Whyte, G., Mahon, N., McKenna, W.J. Physiologic limits of left ventricular hypertrophy in elite junior athletes: Relevance to differential diagnosis of athlete's heart and hypertrophic cardiomyopathy. *J Am Col Cardiol.* 2002; 40(8):1431-1436.

Sharma, S., Whyte, G., Elliott, P.M., Padula, M., Kaushal, R., Mahon, N., McKenna, W.J. Electrocardiographic changes in 1000 highly trained elite athletes. *Br J Sports Med.* 1999;30: 319-324.

Sharma, S., Whyte, G., McKenna, W.J. Sudden cardiac death in young athletes —Fact or fiction? *Br J Sports Med.* 1997;31(4): 269-276.

Smith, D.M., Kovan, J.R., Rich, B.S., et al. *Preparticipation Physical Evaluation.* 2nd ed. Minneapolis: McGraw-Hill; 1997.

Thiene G, Basso C, Corrado D. Sudden death in the young and in the athlete: causes, mechanisms and prevention. *Cardiologia.* 1999 Dec;44 Suppl 1(Pt 1): 415-21.

Thompson, P., Funk, E., Charlton, R., Sturner, W. The incidence of death during jogging in Rhode Island from 1975 through 1980. *JAMA.* 1982;247:2535-2538.

Van Camp, S.P., Bloor, C.M., Mueller, F.O., et al. Non-traumatic sports death in high school and college athletes. *Med Sci Sports Exerc.* 1995;27:641-647.

Whyte, G., George, K., Middleton, N., Shave, R., Nevill, A. Training induced changes in maximal heart rate. *Int J Sports Med.* 2008;29(2):129-133.

Whyte, G.P., George, K., Neville, A., Shave, R., Sharma, S., McKenna, W.J. Left ventricular morphology and function in female athletes: A meta-analysis. *Int J Sports Med.* 2004;25:380-383.

Whyte, G., George, K., Sharma, S., Firoozi, S., Stephens, N., Senior, R., McKenna, W.J. The upper limit of physiologic cardiac hypertrophy in elite male and female athletes: The British experience. *Eur J Appl Physiol.* 2004;31:592-597.

Whyte, G., George, K., Sharma, S., McKenna, W.J. Exercise gas exchange response in the differentiation of physiologic and pathologic left ventricular hypertrophy. *Med Sci Sports Exerc.* 1999;31(9):1237-1241.

Whyte, G., Stephens, N., Budgett, R., Sharma, S., Shave, R., McKenna, W. Exercise induced vasodepressor syncope in an elite rower: A treatment dilemma. *Br J Sports Med.* 2004a;38(1):84-85.

Whyte, G., Stephens, N., Budgett, R., Sharma, S., Shave, R., McKenna, W. Spontaneous atrial fibrillation in a freestyle skier. *Br J Sports Med.* 2004b;38(2):230-232.

Whyte, G., Stephens, S., Senior, S., Peters, N., O'Hanlon, R., Sharma, S. Differentiation of RVOT-VT and ARVC in an elite athlete. *Med Sci Sports Exerc.* 2008;40(8):1357-1361.

Wilson, M.G., Basavarajaiah, S., Whyte, G.P., Cox, S., Loosemore, M., Sharma, S. Efficacy of personal symptom and family history questionnaires when screening for inherited cardiac patholo-gies: The role of electrocardiography. *Br J Sports Med.* 2008;42(3):207-211.

Chapter 7

ACC/AHA/ESC guidelines for the management of patients with atrial fibrillation. Eur Heart Journal. 2001;22:1852-1923.

Ainsworth, C., Skanes, A., Klein, G., Gula, L., Yee, R., Krahn, D. Differentiating arrhythmogenic right ventricular cardiomyopathy from right ventricular outflow tract ventricular tachycardia using multi-lead QRS duration and axis. Heart Rhythm. 2006;3:416-423.

Basavarajaiah, S., Shah, A., Sharma, S. Sudden cardiac death in young athletes. Heart. 2007; 93(3):287-289.

Biffi, A., Maron, B.J., Verdile, L., et al. Impact of physical deconditioning on ventricular tachyarrhythmias in trained athletes. J Am Coll Cardiol. 2004;44:1053-1058.

Biffi, A., Pelliccia, A., Verdile, L, Fernando, F., Spataro, A., Caselli, S., Santini, M., Maron, B.J. Long-term clinical significance of frequent and complex ventricular tachyarrhythmias in trained athletes. *J Am Coll Cardiol.* 2002;40(3):446-452.

Bjornstad, H., Corrado, D., Pelliccia, A. Prevention of sudden death in young athletes: A milestone in the history of sports cardiology. *Eur J Cardiovasc Prev Rehabil.* 2006;13(6):857-858.

Bruce, R.A. Exercise testing of patients with coronary heart disease. Principles and normal standards for evaluation. *Ann Clin Res.* 1971;(6):323-332.

Buxton, A., Waxman, H., Marchlinski, F. Right ventricular tachycardia: Clinical electrophysiologic characteristics. *Circulation.* 1983;68:917-927.

Chandrasekaran, B., Kurbaan, A. Myocardial infarction with angiographically normal coronary arteries. *J Royal Soc Med.* 2002;95:398-400.

Colivicchi, F., Ammirati, F., Biffi, A., Verdile, L., Pelliccia, A., Santini, M. Exercise-related syncope in young competitive athletes without evidence of structural heart disease: Clinical presentation and long-term outcome. *Eur Heart J.* 2002;23:1125-1130.

Cowan, J.C., Yusoff, K., Moore, M., Amos, P.A., Gold, A.E., Bourke, J.P., Transuphaswadikul, S., Campbell, R.W. Importance of lead selection in QT interval measurement. *Am J Cardiol.* 1988:61(1):83-87.

Curran, M.E., Splawski, I., Timothy, K.W., Vincent, G.M., Green, E.D., Keating, M.T. A molecular basis for cardiac arrhythmia: HERG mutations cause long QT syndrome. *Cell.* 1995;80(5):795-803.

Dawson, E., George, K. P., Shave, R., Whyte, G., Ball, D. Does the heart fatigue subsequent to prolonged exercise in humans? *Sports Med.* 2003;3:365-380.

Douglas, P.S., O'Toole, M.L., Hiller, D.B., Hackney, K., Reichek, N. Cardiac fatigue after prolonged exercise. *Circ.* 1987;76:1206-1213.

Funck-Brentano, C., Jaillon, P. Rate-corrected QT interval: Techniques and limitations. *Am J Cardiol.* 1993;72:17B.

Furlanello, F., Bertoldi, A., Dallago, M. Atrial fibrillation in elite athletes. *J Cardiovasc Electrophysiol.* 1998:9:S63-S68.

Gemayel, C., Pelliccia, A., Thompson P. Arrhythmogenic right ventricular cardiomyopathy. *J Am Coll Cardiol.* 2001;38:1773-1781.

Globits, S., Kreiner, G., Frank, H. Significance of morphological abnormalities detected by MRI in patients undergoing successful ablation of right ventricular outflow tract tachycardia. *Circulation.* 1997;96:2633-2640.

Grubb, B., Temesy-Armos, P., Samoil, D., Wolfe, D., Hahn, H., Elliott, L. Tilt table testing in the evaluation and management of athletes with recurrent exercise-induced syncope. *Med Sci Sports Exerc.* 1993;25:24-28.

Haissaguerre, M., Jais, P., Shah, D. Spontaneous initiation of atrial fibrillation by ectopic beats originating in the pulmonary veins. *N Eng J Med,.* 1998;339: 659-666.

Heidbuchel, H., Hoogsteen, J., Fagard, R., Vanhees, L., Ector, H., Willems, R. High prevalence of right ventricular involvement in endurance athletes with ventricular arrhythmias. Role of an electrophysiologic study in risk stratification. *Eur Heart J.* 2003;24:1473-1480.

Hirimoto, M., Akino, M., Takenaka, T., Igarashi, K., Inoue, H., Kawakami Y. Evolution of left ventricular involvement in arrhythmogenic right ventricular cardiomyopathy. *Cardiology.* 2000;93:197-200.

Kalaga, R., Malik, A., Thompson, P. Exercise-related spontaneous coronary artery dissection: A case report and literature review. *Med Sci Sports Exerc.* 2007;39:1218-1220.

Kardasz, I., De Caterina, R. Myocardial infarction with normal coronary arteries: A conundrum with multiple aetiologies and variable prognosis: An update. *J Intern Med.* 2007;261:330-348.

Klein, L., Shigh, H., Hackett F. Radiofrequency catheter ablation of ventricular tachycardia in patients without structural heart disease. *Circulation.* 1992;85:1666-1674.

Leclercq, J., Rosengarten, M., Attuel P. Idiopathic ventricular extrasystole: Right ventricular parasystole not protected from the sinus rhythm? *Arch Mal Coeur Vaiss.* 1981;74:1249-1261.

Levine, B., Lane, L., Buckey, J., Friedman, D., Blomqvist, G. Left ventricular pressure-volume and Frank-Starling relations in endurance athletes: Implications for orthostatic tolerance and exercise performance. *Circulation.* 1991;84:1016-1023.

Link, M., Homoud, M., Wang, P., Estes, M. Cardiac arrhythmias in the athlete. *Cardiol Rev.* 2001;9:21-30.

Malik, M., Färbom, P., Batchvarov, V., Hnatkova, K., Camm, A.J. Relation between QT and RR intervals is highly individual among healthy subjects: Implications for heart rate correction of the QT interval. *Heart.* 2002 Mar: 87(3): 220-228.

Monnig, G., Eckardt, L., Wedekind, H., Haverkamp, W., Gerss, J. Milberg, P., Wasmer, K., Kirchof, P., Assman, G., Breithardt, G., Schulze-Bah. Electrocardiographic risk stratification in families with congenital long QT syndrome. *Eur Heart J.* 2006;27:2074-2080.

Moss A. Measurement of the QT interval and the risk associated with QTc prolongation: A review. *Am J Cardiol.* 1993;72:23B.

Moss, A.J. Prolonged QT syndromes. *JAMA.* 1986;256:2985-2987.

Napolitano, C., Priori, S.G., Schwartz, P.J., Bloise, R., Ronchetti, E., Nastoli, J., Bottelli, G., Cerrone, M., Leonardi, S. Genetic testing in the long QT syndrome: Development and validation of an efficient approach to genotyping in clinical practice. *JAMA.* 2005;294:2975.

Newby, N.L. Markers of cardiac ischemia, injury, and inflammation. *Prog Cardiovas Dis.* 2004;46:404-416.

Nistri, S., Thiene, G., Basso, C., Corrado, D., Vitolo, A., Maron, B.J. Screening for hypertrophic cardiomyopathy in a young male military population. *Am J Cardiol.* 2003;91(8):1021-1023.

O'Connor, F., Pyne, S., Brennan, F., Adirim, T. Exercise associated collapse: An algorithmic approach to race day management. *Am J Med Sports.* 2003;5:212-217.

Pelliccia, A., Corrado, D., Bjornstad, H.H., Panhuyzen-Goedkoop, N., Urhausen, A., Carre, F., Anastasakis, A., Vanhees, L., Arbustini, E., Priori, S. Recommendations for participation in competitive sport and leisure-time physical activity in individuals with cardiomyopathies, myocarditis and pericarditis. *Eur J Cardiovasc Prev Rehabil.* 2006;13(6):876-885.

Pelliccia, A., Maron, B.J., Culasso, F., Di Paolo, F.M., Spataro, A., Biffi, A., Caselli, G., Piovano, P. Clinical significance of abnormal electrocardiographic patterns in trained athletes. *Circulation.* 2000:18;102(3):278-284.

Priori, S.G., Schwartz, P.J., Napolitano, C., Bloise, R., Ronchetti, E., Grillo, M., Vicentini, A., Spazzolini, C., Nastoli, J., Bottelli, G., Folli, R., Cappelletti, D. Risk stratification in the long-QT syndrome. *N Engl J Med.* 2003;348(19):1866-1874.

Quaglini, S., Rognoni, C., Spazzolini, C., Priori, S.G., Mannarino, S., Schwartz, P.J. Cost-effectiveness of neonatal ECG screening for the long QT syndrome. *Eur Heart J.* 2006;27:1824.

Raymond, R., Lynch, J., Underwood, D., Leatherman, J., Razavi, A. Acute myocardial infarction and normal coronary angiography 10 years clinical and risk analysis in 74 patients. *J Am Coll Cardiol.* 1988;11:471-477.

Resuscitation Council (UK). *Advanced Life Support Course Provider Manual.* 4th ed. Resuscitation Council (UK); 2004.

Richter, S., Hohnloser M. Elecrocardiographic hallmark of arrhythmogenic right ventricular cardiomyopathy. *J Cardiovasc Electrophysiol.* 2006;17:563-564.

Robisek, F. Myocardial infarction with angiographically normal coronary arteries. *J Royal Soc Med.* 2002;95:528.

Sallis, R. Collapse in the endurance athlete. *Sports Sci Exchange.* 2004;12:1-5.

Sastry, S., Riding, G., Morris, J., Taberner, D., Cherry, N., Heagerty, A., McCollum, C. Young adult myocardial infarction and ischemic stroke: The role of paradoxical embolism and thrombophilia (The AMIS study). *J Am Coll Cardiol.* 2006;15:686-691.

Schmidt, A., Azevedo, C.F., Cheng, A. Infarct tissue heterogeneity by magnetic resonance imaging identifies enhanced cardiac arrhythmia susceptibility in patients with left ventricular dysfunction. *Circulation.* 2007;115:2006-2014.

Schwartz, P.J. The long QT syndrome. *Curr Probl Cardiol.* 1997;22:297.

Schwartz, P.J., Periti, M., Malliani, A. The long QT syndrome. *Am Heart J.* 1975;89:378-390.

Sen-Chowdhry, S., Lowe, M., Sporton, S., McKenna, W.J. Arrhythmogenic right ventricular cardiomyopathy: Clinical presentation, diagnosis, and management. *Am J Med.* 2004;117:685-695.

Sharma, S., Whyte, G., Elliott, P.M., Padula, M., Kaushal, R., Mahon, N., McKenna, W.J. Electrocardiographic changes in 1000 highly trained elite athletes. *Br J Sports Med.* 1999;30:319-324.

Shave, R.E., Whyte, G.P., George, K., Gaze, D.C., Collinson, P.O. Prolonged exercise should be considered alongside typical symptoms of AMI when evaluating elevations in cardiac troponin T. *Heart.* 2005;I91:1219-1220.

Sneddon, J., Scalia, G., Eard, D., McKenna, W.J., Camm, J., Frenneaux, M. Exercise induced vasodepressor syncope. *Br Heart J.* 1994;71:554-557.

Srivathsan, K., Lester, S., Appleton, C., Scott, L., Munger T. Ventricular tachycardia in the absence of structural heart disease. *Indian Pacing and Electrophysiology Journal.* 2005;5:106-121.

Stegmann, J., Buster, A., Brock, D. Influences of fitness on the blood pressure control system in man. *Aerospace Med.* 1974;45:45-48.

Sullivan, M., Martinez, C., Gallagher, E. Atrial fibrillation and anabolic steroids. *J Emerg Med.* 1999;17:851-857.

Thiene, G., Bassos, C., Corrado D. Is prevention of sudden death in young athletes feasible? *Cardiologica.* 1999;44:497-505.

Thiene, G., Nava, A., Corrado, D., Rossi, L,. Pennelli N. Right ventricular cardiomyopathy and sudden death in young people. *N Engl J Med.* 1988;318:129-133.

Tsatsopoulou, A., Protonotarios, N., McKenna W.J. Arrhythmogenic right ventricular dysplasia, a cell adhesion cardiomyopathy: Insights into disease pathogenesis from preliminary genotype-phenotype assessment. *Heart.* 2006;92:1720-1723.

Tun, A., Khan, I. Myocardial infarction with normal coronary arteries: The pathologic and clinical perspectives. *Angiology.* 2001;52:299-304.

Van Baak, M. β-adrenoreceptor blockade and exercise: An update. *Sports Medicine.* 1988;4:209-225.

Wang, Q., Curran, M.E., Splawski, I., Burn, T.C., Millholland, J.M., VanRaay, T.J., Shen, J., Timothy, K.W., Vincent, G.M., de Jager, T., Schwartz, P.J., Toubin, J.A., Moss, A.J., Atkinson, D.L., Landes, G.M., Connors, T.D., Keating, M.T. Positional cloning of a novel potassium channel gene: KVLQT1 mutations cause cardiac arrhythmias. *Nat Genet.* 1996;12(1):17-23.

Wang, Q., Shen, J., Splawski, I., Atkinson, D., Li, Z., Robinson, J.L., Moss, A.J., Towbin, J.A., Keating, M.T. SCN5A mutations associated with an inherited cardiac arrhythmia, long QT syndrome. *Cell.* 1995;80(5):805-811.

Whyte G. P., George, K., Sharma, S., Lumley, S., Gates, P., Prasad, K., McKenna, W. J. Cardiac fatigue following prolonged endurance exercise of differing distances. *Med Sci Sports Exerc.* 2000;32:1067-1072.

Whyte, G., Sheppard, M., George, K., Shave, R., Wilson, M., Stephens, N., Senior, R., Sharma, S.

Arrhythmias and the athlete: Mechanisms and clinical significance. *Eur Heart J.* 2007;28(11):1399-1401.

Widimsky, P., Stellova, B., Groch, L., Aschermann, M., Branny, M., Zelizko, M., Stasek, J., Formanek, P. PRAGUE Study Group Investigators. *Can J Cardiol.* 2006;22:1147-1152.

Yarlagadda, R.K., Iwai, S., Stein, K.M., Markowitz, S.M., Shah, B.K., Cheung, J.W., Tan, V., Lerman, B.B., Mittal, S. Reversal of cardiomyopathy in patients with repetitive monomorphic ventricular ectopy originating from the right ventricular outflow tract. *Circulation.* 2005;112(8): 1092-1097.

Zareba, W., Moss, A.J., Schwartz, P.J., Vincent, G.M., Robinson, J.L., Priori, S.G., Benhorin, J., Locati, E.H., Towbin, J.A., Keating, M.T., Lehmann, M.H., Hall, W.J. Influence of genotype on the clinical course of the long-QT syndrome. International Long-QT Syndrome Registry Research Group. *N Engl J Med.* 1998;339(14): 960-965.

Index

Note: The italicized *f* and *t* following page numbers refer to figures or tables, respectively.

About the Authors

Photo courtesy of Greg Whyte

Greg Whyte, PhD, FACSM, is a professor of applied sport and exercise science at Liverpool John Moores University in London. As one of British sport's foremost research scientists, Whyte has worked as a consultant physiologist in a large number of Olympic and professional sports and is currently the science consultant for England FA (World Cup 2010), the Commonwealth Games Committee for England (Delhi 2010), and British Rowing. From 2001 to 2004, Whyte served as the director of research for the British Olympic Association based at the Olympic Medical Institute, where he established the Centre for Sports Cardiology, which is dedicated to the investigation and treatment of sport-related cardiac issues. Whyte is now the director of the Centre for Sports Cardiology at the Centre for Health and Human Performance.

A former international modern pentathlete, Whyte competed in two Olympic Games and won European bronze and World Championship silver medals. He studied for his BSc (hons) at Brunel University, completed his MSc in human performance at Frostburg State University in the United States, and completed his PhD at St George's Hospital Medical School and the University of Wolverhampton, where he was research coordinator. Whyte is a fellow of the American College of Sports Medicine (FACSM) and was chairman of the charity, Cardiac Risk in the Young (CRY), from 1999 to 2009.

Photo courtesy of Sanjay Sharma

Sanjay Sharma, BSc (Hons), MD, FRCP (UK), FESC, was appointed consultant cardiologist and physician at University Hospital Lewisham and honorary senior lecturer in cardiology at Kings College Hospital London in 2001. In 2006 he took up the post of director of heart muscle diseases at Kings College in London and became professor of cardiology at St George's University of London in 2009. Sharma is medical director for Virgin London Marathon, consultant cardiologist for the CRY sports cardiology clinic at St George's Hospital, and cardiologist for the English Institute of Sport, the British Rugby League, and the British Lawn Tennis Association.

Sharma's interests include cardiovascular adaptation in athletes, sudden cardiac death in the young, and heart muscle diseases, for which he has an international reputation and has published over 100 scientific articles, including original papers in highly rated peer-reviewed journals. Sharma was awarded the status of fellow of the European Society of Cardiology and elected as a nucleus member of the Sport Cardiology section of the European Association of Cardiovascular Disease Prevention and Rehabilitation in 2008. Sharma leads the CRY screening program, which is the largest of its kind in the UK. Sharma has an active interest in medical education and is the lead tutor for the international teaching faculty for the Royal College of Physicians.